I0034504
```

James R. Halladay, R J Del Vecchio
**Bonding of Elastomers**
```

Also of Interest

Rubber.
Science and Technology
Princi, 2019
ISBN 978-3-11-064031-1, e-ISBN 978-3-11-064032-8

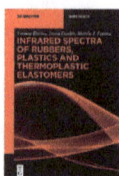

Infrared Spectra of Rubbers, Plastics and Thermoplastic Elastomers.
Davies, Davies, Forrest, 2019
ISBN 978-3-11-064408-1, e-ISBN 978-3-11-064575-0

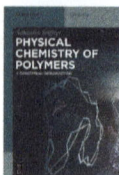

Physical Chemistry of Polymers.
A Conceptual Introduction
Seiffert, 2020
ISBN 978-3-11-067280-0, e-ISBN 978-3-11-067281-7

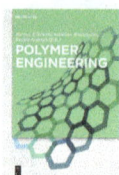

Polymer Engineering.
Tylkowski, Wieszczycka, Jastrzab (Eds.), 2017
ISBN 978-3-11-046828-1, e-ISBN 978-3-11-046974-5

Porous Polymer Chemistry.
Synthesis and Applications
Yavuz, 2021
ISBN 978-3-11-049465-5, e-ISBN 978-3-11-049468-6

James R. Halladay, R J Del Vecchio

Bonding of Elastomers

A Practical Guide

DE GRUYTER

Authors
James R. Halladay
1530 Taylor Ridge Court
Erie, PA 16505
USA
jrhalla@aol.com

R. J. Del Vecchio
31 Brickyard Rd.
Asheville, NC 28806
USA
TechConsultServ@Juno.com

ISBN 978-3-11-065649-7
e-ISBN (PDF) 978-3-11-065899-6
e-ISBN (EPUB) 978-3-11-065660-2

Library of Congress Control Number: 2020939129

Bibliographic information published by the Deutsche Nationalbibliothek
The Deutsche Nationalbibliothek lists this publication in the Deutsche Nationalbibliografie;
detailed bibliographic data are available on the Internet at http://dnb.dnb.de.

© 2020 Walter de Gruyter GmbH, Berlin/Boston
Cover image: James R. Halladay
Typesetting: Integra Software Services Pvt. Ltd.
Printing and binding: CPI books GmbH, Leck

www.degruyter.com

Preface

Why is bonding important?

The 3 original uses of tree latex in South America were for making balls for games, waterproof shoes, and waterproof fabric, and those products remain in wide use today. The first 2 have nothing to do with bonding but for the homespun fabric to be useful for raincoats and so on meant the dried rubber coating had to stick to it. Fortunately, natural rubber latex wets out nicely on organic fibers and naturally forms a reasonably stable bond to them. A raincoat that has the barrier layer to water that deteriorates and falls off would not be useful to anyone.

Until vulcanization was discovered, rubber products were in gum form but were still useful in various applications. Making rain gear progressed from drying latex-wet fabric over wood fires to dissolving the rubber in a solvent to form a paste or viscous liquid and then coating the fabric in a continuous operation therewith and drying it. This would leave a somewhat controlled layer of rubber on and to some extent, depending on weave, into the fabric, to produce a useable product.

Engineers could soon see many possible uses for rubber products that were somehow bonded to a solid, but very few applications were developed using simple gum materials.

Once vulcanization was discovered, the use of rubber started to expand rapidly in more and more applications, and the capacity to bond somewhat to fibers became useful in the fledgling tire industry. But what engineers really wanted was the ability to bond to metals.

Experimentation with sulfur vulcanization systems led to the discovery of ebonite, the hard, plastic-like solid that had its own set of applications. It could be made in sheets, blocks, and rods and drilled, machined, and used in applications like battery cases, bowling balls, and small items like combs, pens, and jewelry. Further experimentation demonstrated that ebonite had a capacity to bond to metal when vulcanized against a clean surface and that led to the long and still ongoing technical field of bonding rubbers to solid surfaces. (See more on this in Chapter 2.)

Just getting a piece of rubber to stick to some solid object, however weakly and however briefly, is not what makes for really serviceable products. One of the most common applications of bonding rubber is in modern tires, where the strength and functionality of the tire hinges on the use of woven steel belts as a critical part of the overall complex structure of a tire. For the tire to have any reasonable working lifetime, those metal belts must be bonded strongly and lastingly in the rubber mass that makes up most of the tire. The life of that bond must be longer than the time taken for the tire tread to wear out.

https://doi.org/10.1515/9783110658996-202

If that bond fails, then there can be and often is a catastrophic failure resulting in major damage, injuries, and death. There have been 2 famous cases of tire design failures that led to extremely unfortunate results.

Many engineered rubber products involved bonded parts, for instance the myriad types of mounts that are used in cars, trucks, aircraft, heavy construction equipment, and more. If the bonds do not last through years of use, often through a wide range of environmental conditions, equipment will function badly or break down, and replacement of the broken parts can be hugely expensive. There are bonded rubber parts used in places most don't think of, like down hole in oil wells where conditions are incredibly aggressive, and replacing such parts if they fail can cost tens of thousands of dollars.

Cost effectiveness of many products depends on both long-lasting rubber materials and long-lasting, durable bonds. Conveyor belts for materials like coal or hot slag in refineries are complex composite structures, like tires, and so are other products such as snowmobile tracks and other kinds of vehicle tracks. These are not simple products to manufacture and have correspondingly high costs, and the longer the lifetime they can provide, the more successful they become as products. But in any of them the lifetime of the bond can dictate the functional life of the product, so again bond strength and durability under very taxing conditions of use become of paramount importance.

The demands for strong bonds that last for very long times under challenging conditions have been increasing for decades. With the use of more specialized polymers that are more difficult to bond, plus the use of more substrates that present their own difficulties in allowing bonds to form, the technical field of rubber bonding has become more complicated. This book is intended to give those whose work involves producing bonded rubber products a thorough grounding in this methodology.

Contents

1 Adhesion theory

Adhesives are essential to modern day mechanical devices. The transportation industry, both aerospace and industrial, relies on adhesives for virtually all products. Elastomers are used in many applications as flexible barriers like seals, O-rings, and diaphragms, and for energy management applications like tires, vibration isolators, and dampers (Figure 1.1). Many of these applications require that the elastomer be bonded or adhered to a substrate, commonly metal, textiles, or rigid plastic, and their ability to function would be impossible without robust rubber-to-substrate bonds. Tires, engine mounts, elastomeric bearings, vibration isolators, dampers, and the solid fuel rocket engines for the space shuttle are but a few examples.

Much literature has been published on the history and technology of bonding rubber to metal [1–7]. First, we will define some terms commonly used throughout the industry. ASTM D907 defines adhesion as a state in which two surfaces are held together by interfacial forces which may consist of valence forces or interlocking forces or both. The term "adhesion" refers to the strength of forces holding together two separate surfaces of the same or different materials. Thus, an adhesive is a substance that is able to hold materials together by surface attachment. "Cohesion" refers to the strength of a single material in holding itself together. In addition to having good adhesion to a substrate, an adhesive must have sufficient cohesive strength to support load. A primer is a substance applied to a substrate to generate good adhesion to the substrate, and it must also be capable of interacting with an adhesive topcoat.

Adhesion can be achieved by any combination of factors including mechanical interlocking, molecular interdiffusion, electrostatic forces, adsorption, and chemical bond formation. A number of well-known attractive forces (ionic, covalent, metallic, hydrogen, van der Waals) ensure the cohesion and adhesion of solids[8].

The forces of attraction between atoms and molecules can be divided into interatomic bonds and secondary bonds. Interatomic bonds are the strongest bonds and include ionic bonds, covalent bonds, and intermetallic bonds. Ionic bonding occurs when one of the atoms is negative (has an extra electron) and another is positive (has lost an electron). Ionic bonds are the strongest bonds, greater than 5×10^{-4} dynes/bond, and they create a strong Coulomb attraction. A good example of ionic bonding is sodium chloride salt, NaCl (Figure 1.2).

https://doi.org/10.1515/9783110658996-001

Figure 1.1: Cross section of laminated helicopter rotor bearing (made by Lord Corporation).

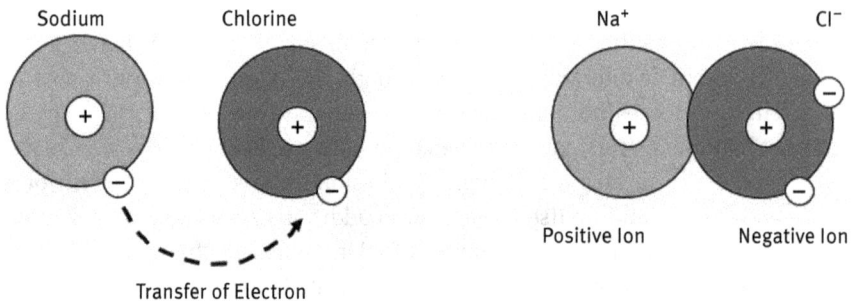

Figure 1.2: Ionic bonding.

Covalent bonding occurs when electrons are shared between molecules to saturate the valence. Covalent bonds are slightly less strong than ionic bonds, approximately 5×10^{-4} dynes/bond (Figure 1.3).

In metals, the atoms are ionized, having lost some electrons from the valence band (the electrons are delocalized). These electrons form an electron sea which binds the charged nuclei in place. Intermetallic bonds are only about half as strong as covalent bonds, approximately 2×10^{-4} dynes/bond (Figure 1.4).

Secondary bonds range from weak to very weak forces and include hydrogen bonds, London dispersion forces and Van der Waals forces. Polar molecules such as water have a weak, partial negative charge at one region of the molecule and a partial positive charge elsewhere. The positive and negative regions are attracted to the oppositely charged regions of nearby molecules. Hydrogen bonding is electrostatic because hydrogen has only a single electron

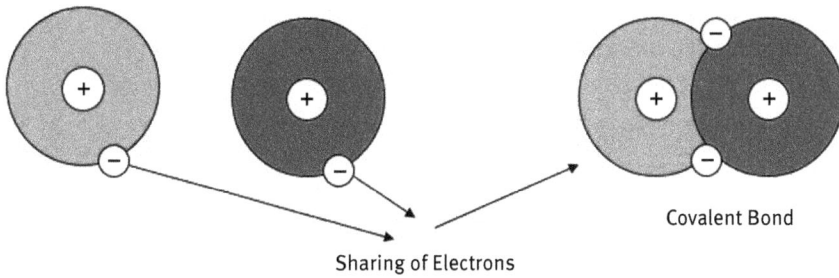

Sharing of Electrons

Covalent Bond

Figure 1.3: Covalent bonds.

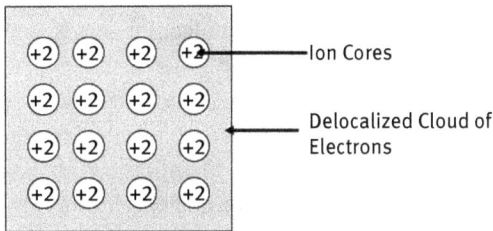

Ion Cores

Delocalized Cloud of Electrons

Figure 1.4: Intermetallic bonds.

in the 1s orbital and cannot form covalent bonds. Hydrogen bonding involves molecules with OH, NH or FH groups and hydrogen bonds are only 5–10% of the strength of covalent bonds, approximately 6×10^{-5} dynes/bond (Figure 1.5).

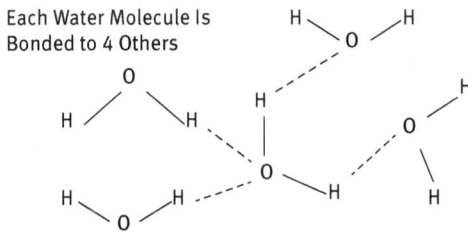

Each Water Molecule Is Bonded to 4 Others

Figure 1.5: Hydrogen bonds.

Van der Waals forces, also known as dipole–dipole forces, are caused by the electrostatic attraction between polar molecules. The magnitude of force increases with the number of electrons per molecule. They are much, much weaker than chemical bonds and much weaker than hydrogen bonding, approximately 2×10^{-6} dynes/bond (Figure 1.6).

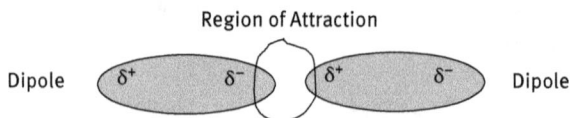

Region of Attraction

Dipole δ^+ δ^- δ^+ δ^- Dipole

Figure 1.6: Van der Waals Forces.

London dispersion forces describe the transitory electrostatic attraction that results when the electrons of two adjacent atoms occupy positions that make the atoms form temporary dipoles. These are weak intermolecular forces that arise from the interactive forces between instantaneous dipoles in molecules without permanent dipole moments (Figure 1.7). These forces dominate the interaction of nonpolar molecules, and are significant in polar molecules. London dispersion forces are also known as "dispersion forces", "London forces", or "instantaneous dipole–induced dipole forces". The strength of London dispersion forces is proportional to the polarizability of the molecule, which in turn depends on the total number of electrons and the area over which they are spread. They are approximately 2×10^{-6} dynes/bond.

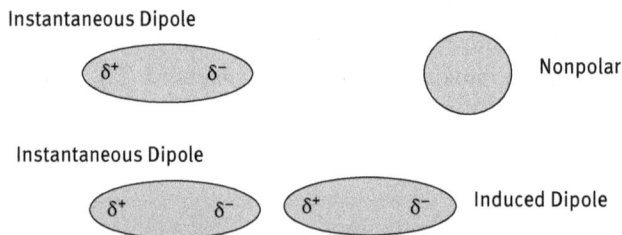

Instantaneous Dipole

δ^+ δ^- Nonpolar

Instantaneous Dipole

δ^+ δ^- δ^+ δ^- Induced Dipole

Figure 1.7: London dispersion forces.

While permanent chemical bonds like covalent bonds are certainly desirable, covalent bonds alone cannot provide the strength we expect in a bonded specimen. The maximum theoretical strength calculated from covalent bonds is well below the strength obtained in bonded systems. Thus, the weak secondary bonds are also extremely important to the formation of a robust adhesive bond. Molecular weight and branching are important in an adhesive as well. Increasing molecular weight gives increasing strength as one progresses from a monomer (generally a liquid or gas) to an oligomer to a polymer, for example, going from ethylene to n-paraffins to polyethylene. Likewise, increasing molecular branching increases strength as one progresses from liquids to gels to rigid solids.

Adhesion is increased by molecular interdiffusion which gives interlocking on a molecular level. Examples include autohesion (adhesion of a polymer to itself), polymer welding, and miscible polymers which exhibit diffusion between the dissimilar phases. Anyone who has laid one piece of rubbery polymer like polybutadiene or polyisobutylene on another piece knows that they become increasingly difficult to separate over time. Molecular interdiffusion of the polymers is affected by time, temperature, and pressure as shown in Figures 1.8–1.10.

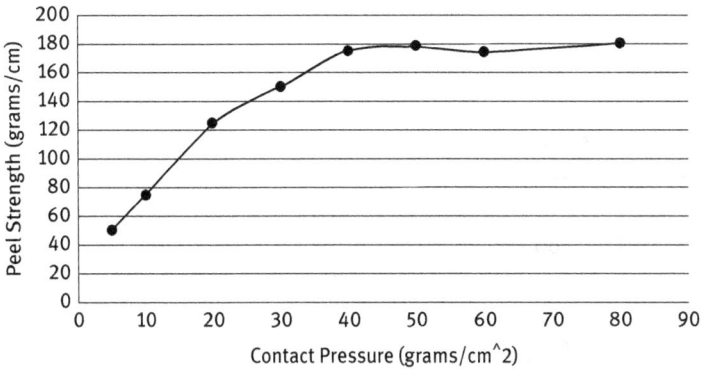

Figure 1.8: Effect of contact pressure on polyisobutylene autohesion.

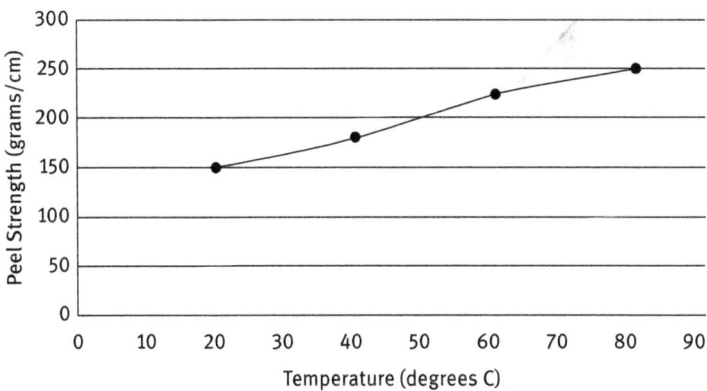

Figure 1.9: Effect of contact temperature on polyisobutylene autohesion.

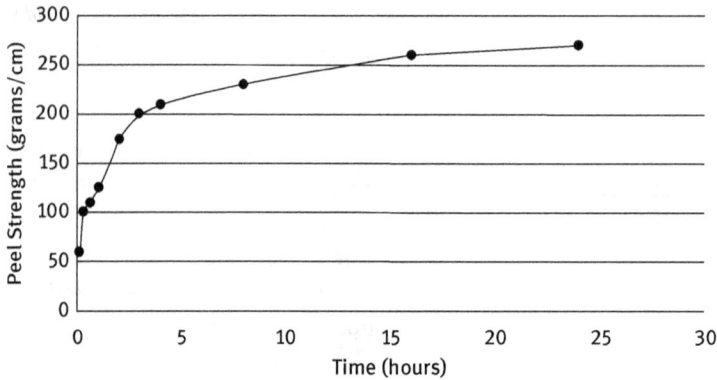

Figure 1.10: Effect of contact time on polyisobutylene autohesion.

Wetting is important to adhesive chemistry as well. The deposition of a coating on a solid generates new interfaces between dissimilar materials and involves considerations of wettability. Wetting is a basic or fundamental concept of adhesion but it is also important in the practical performance of adhesives. To really work well, an applied liquid adhesive should have intimate contact with each surface and spread freely on both. The wetting of a solid by a liquid is characterized in terms of the angle of contact that the liquid makes on the solid (Figure 1.11) [9]. The contact angle of wetting is denoted as θ (theta). For totally free spreading, θ equals zero, but θ is usually finite because forces of attraction among the liquid's molecules are greater than forces of attraction of the liquid molecules to those of the solid.

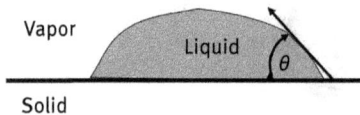

Figure 1.11: Liquid wetting of a solid surface.

Surface tension or surface energy is expressed by symbol y (gamma).

$$\gamma_{sv} = \gamma_{sl} + \gamma_{lv}(cos\theta)$$

where
y_{sv} = solid/vapor surface tension
y_{sl} = solid/liquid surface tension
y_{lv} = liquid/vapor surface tension

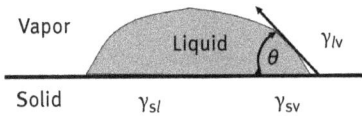

Figure 1.12: Surface tensions.

Spreading pressure (S_{sl}) is a measure of the ability of the liquid to wet and spread on a solid surface and spreading pressure is related to the surface tensions:

$$\text{Spreading Pressure}: \boldsymbol{S_{sl}} = \gamma_{sv} - \gamma_{sl} - \gamma_{lv}$$

An example of the importance of surface tension can be seen in the following example. A mixed liquid epoxy adhesive is applied to the surface of polyethylene. After cure, the cured epoxy will exhibit little or no adhesion to the polyethylene. But molten polyethylene applied to surface of cured epoxy will show fairly strong adhesion. That is because, in the first case, the surface tension of the liquid epoxy is high and the surface tension of the solid polyethylene is low. However, in the second case, the surface tension of the liquid polyethylene is lower than the surface tension of the cured epoxy.

Water has very poor ability to wet the surface of rubber. Water droplets on a cured rubber surface exhibit a high contact angle with θ being around 90 degrees in Figure 1.13.

Figure 1.13: Water droplets on cured rubber.

On aluminum, the contact angle θ shows that the solvent methyl isobutyl ketone (MIBK) is better for wetting the aluminum surface than water (Figure 1.14).

Surface preparation can change the surface energy of both elastomers and metals. For a strong adhesive bond, the surfaces to be bonded must be clean and should have higher surface energy than the adhesive. For elastomers, cleaning the surface by abrasion or solvent wiping will generally not be sufficient unless

Figure 1.14: MIBK (right) and water (left) droplets on aluminum, top view and side view.

the polarity is high (like neoprene CR or nitrile NBR). Cyclization using concentrated sulfuric acid or chlorination using trichloroisocyanuric acid (TCICA) is often employed. Metal surfaces are often prepared using grit blasting, chemical degreasing, chemical etches, conversion coatings, brass plating, or combinations thereof (more in a later chapter).

The strength of a bond joint depends on the mode of loading, the dimensions, the elastic properties of bonded components, and the intrinsic strength of the interface(s). Fracture mechanics relate observed breaking load to these variables. Fracture mechanics was originally proposed for brittle fracture of materials like metals and glass. The field has now evolved to be useful for materials that have become locally dissipative and for truly elastic materials. The forces required to break apart brittle solids (e.g., crystals of sodium chloride for example) are much, much less than theory predicts. The actual forces of rupture or disbonding equate roughly with relatively weak London or van der Waals forces of attraction (those that would result from simple adsorption) rather than the covalent bonds that actually exist.

Fracture mechanics explains very low forces of disbonding in terms of structural defects, microcracks, and crack propagation. But fracture energy can be substantially restored by reducing brittleness in materials including adhesives by introducing rubbery ingredients for example. For this reason, most adhesives contain rubbery polymers as part of their formulation.

References

[1] Alstadt, D. M., "Some Fundamental Aspects or Rubber-Metal Adhesion", Rubber World, November, 1955, p 133.

[2] Alstadt, D. M., and Coleman, E. W., Jr., U.S. Patent 2,905,585, to Lord Corporation (September 1959).

[3] Buchan, S., Rubber to Metal Bonding, Palmerton, New York, 1959.

[4] DeCrease, W. M., Rubber Age, 87, 1960, 1013–1019. This publication has been out of print for 30+ years and details are no longer recoverable.

[5] Jazenski, P. J., and Manino, L. G., U.S. Patent 4,119,587, to Lord Corporation (October 1978).

[6] Elliot, D. J., Developments in Rubber Technology, 1, Applied Science, London, 1979, 1–44.

[7] Weih, M. A., Siverling, C. E., and Sexsmith, F. H., Rubber World, 195(5), 29–35 (August 1986).

[8] Lee, L. H., Adhesive Chemistry, Developments and Trends, Plenum Press, New York, 1984, 64.

[9] Lee, Michael, "An Analytical Method for Determining the Surface Energy of Polymers", Published in Adhesive Chemistry, Developments and Trends, Edited by Lieng-Huang Lee, Plenum Press, New York, 1983, pp 95–120.

2 History of rubber-to-metal bonding

The earliest historical methods of attaching rubber to metal involved attaching the rubber by mechanical means or by the use of ebonite. Mechanical attachment alone creates an insecure union. Natural rubber is normally cured using one to four parts per hundred (phr) elemental sulfur. When 25 or more phr is used, the resulting vulcanizate, called ebonite, is a rigid solid more like a modern plastic than a rubber. Because of the extremely high sulfur content, the ebonite can react with some metals to create an adhesive bond. The cure reaction in the rubber is basically addition of sulfur at the double bonds, forming intramolecular ring structures. In ebonite, much of the sulfur is believed to be in the form of intramolecular addition since it is noticeably thermoplastic [1]. Ebonite forms a true bond to the softer sulfur-curable rubber and it also adheres rather strongly to the metal.

Ebonite adhesives were made by dissolving ebonite compound in solvent and applying it to the surface of steel. In some cases, a series of adhesives were made by dissolving increasing percentages of the soft natural rubber elastomer to be bonded in the ebonite solution and applying multiple adhesive coats with decreasing sulfur contents. Either way, the sulfur content is reduced with increasing distance from the metal surface. This creates a decreasing modulus gradient between the rigid steel and the soft elastomer (Figure 2.1).

Bonding with ebonite creates several problems. One significant problem is that the ebonite is thermoplastic and becomes quite weak with moderate temperature exposure. Depending on the amount of sulfur, ebonite based on natural rubber shows a thermoplastic transition temperature between 70 and 80 °C. At sulfur levels between 4 phr and 25 phr, cured natural rubber goes through a transition where it becomes rather leathery and has low strength. Because of the sulfur gradient between the ebonite adhesive and the soft rubber compound, at some point, the sulfur content of the compound must pass through this transition zone. This transition zone weakens the softer rubber in the interfacial region and it reduces the flexibility in that region. Bonding with ebonite also limits the chemistry of rubber formulations that can be successfully bonded using this technique.

Another early method of bonding involved the use of special metal alloys which were capable of reacting and combining with sulfur. The earliest patent for the use of alloys was in Germany in 1904 [2]. Daft patented alloys containing antimony in the United States between 1912 and 1913 [3–6]. He also claimed the use of alloys of copper and zinc with bismuth and arsenic. These alloys were electrically deposited on the metal and the bonds to rubber were formed during the vulcanization process.

https://doi.org/10.1515/9783110658996-002

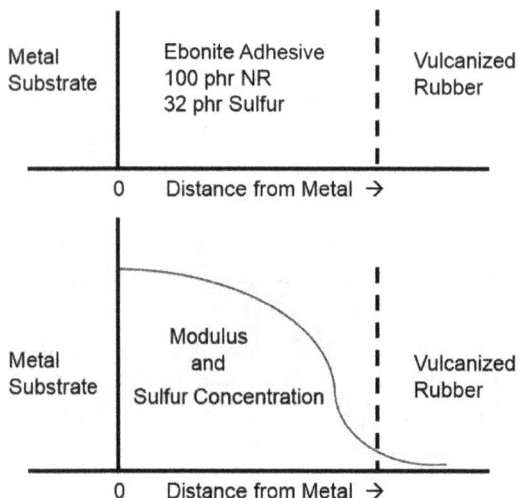

Figure 2.1: Modulus gradient follows sulfur concentration.

In 1862, Sanderson submitted a British patent application for the use of electrodeposited brass as an intermediary for bonding rubber to iron or steel [7]. However, it was not until between 1920 and 1930 that the process of bonding to a galvanic layer of brass (brass plating) was actually commercialized. The brass matrix is typically about 70% copper and adhesion is obtained by virtue of the chemical reaction that occurs between copper and the sulfur curative in the rubber (Figure 2.2). Bonding to the brass plating had the advantage over the ebonite process of not being so heat sensitive.

The galvanic plating process requires a large investment in processing machinery and it is difficult to keep all the variables in the galvanic bath constant. It is somewhat unpredictable and shows a high sensitivity to processing conditions. As with bonding to ebonite, it limits the chemistry of the formulations that can be successfully bonded to only those compounds containing a high sulfur cure system (2 to 4 phr). As a further complication, not all types of brass will bond to rubber equally and it appears that the best ratio of copper to zinc is somewhat compound dependent. In a production environment, consistent results are often hard to obtain without considerable experience and the factors which must be carefully controlled include bath and anode composition, temperature, voltage, current density, time of deposition, and hydrogen-ion concentration. However, the brass plating process has proven quite successful for certain applications such as steel cords for automotive tires.

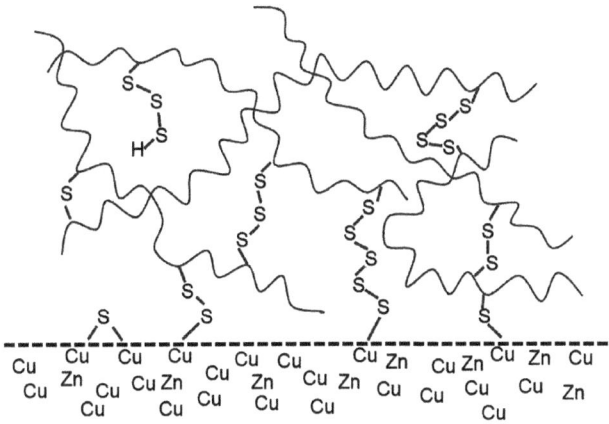

Figure 2.2: Bonding to brass via copper–sulfur–rubber linkages.

The use of cyclized rubber as an adhesive was introduced around 1927. Acid cyclization was accomplished by treating natural rubber dissolved in a solvent with sulfuric acid (Figure 2.3).

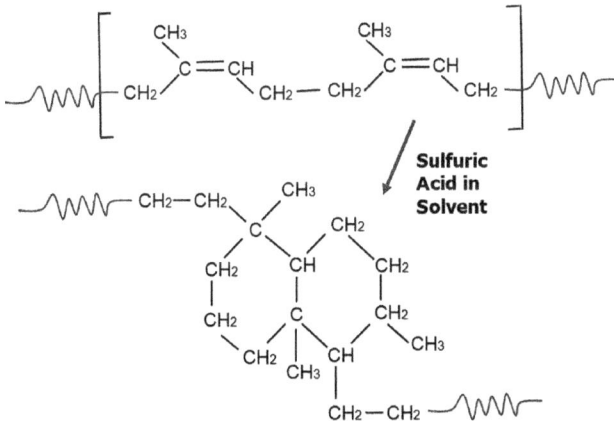

Figure 2.3: Acid cyclization of rubber.

Around 1930, a bonding process using phenolic resins was introduced (Figure 2.4). The phenolic resin resoles chemisorb onto steel and they thermocondense into a three dimensional network.

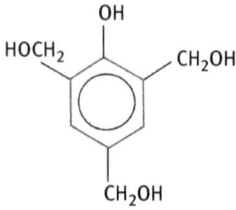

Figure 2.4: Phenolic resin resole.

Chlorinated natural rubber (Figure 2.5) was introduced as an adhesive around 1932.

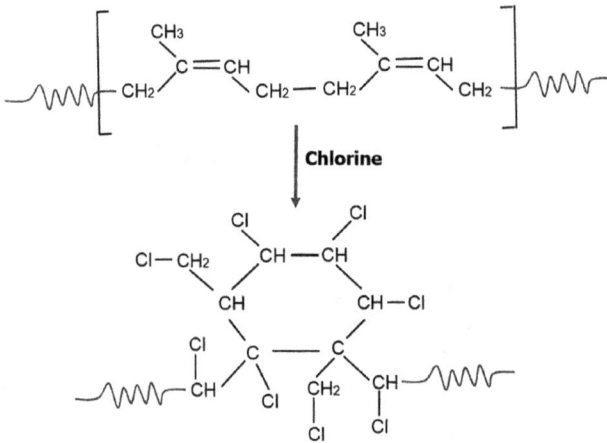

Figure 2.5: Chlorination of natural rubber.

In 1930, Hugh C. Lord received US patent #1,749,824 for bonding natural rubber to steel using a primer similar to a semi-ebonite compound (Table 1.1) and a natural rubber stock.

Table 1.1: Formulations from US patent #1,749,824.

	Primer	Rubber Stock
Natural Rubber	100	100
Zinc Oxide	167	3
Iron Oxide	50	–
Sulfur	17	5
Hexamethylenetetramine	–	1
Lime	4	–

These compounds were dissolved in solvent and blended in proportions to create layers ranging from 100% of the primer formulation to 100% of the rubber stock. In doing so, the concept of a modulus gradient was greatly increased (Figure 2.6).

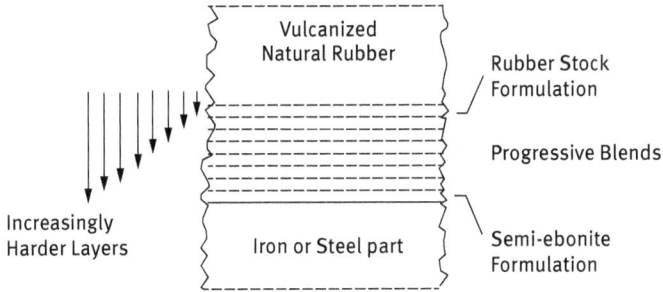

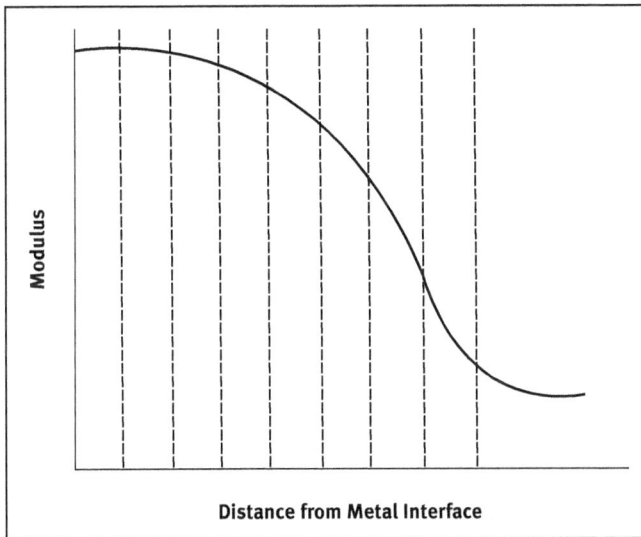

Figure 2.6: From Hugh C. Lord US patent #1,749,824.

Another method of obtaining rubber to metal bonds was discovered in Germany around 1945 and involves the use of isocyanates, in particular triphenylmethane triisocyanate [8, 9] (Figure 2.7).

Polyisocyanates chemisorb strongly on ferrous metal substrates. When applied to clean metal surfaces, they give good primary adhesion between many types of natural and synthetic rubber formulations and a wide variety of metal

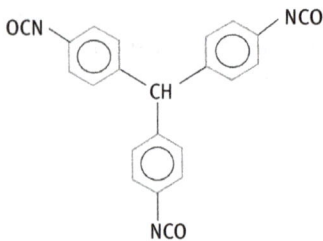

Figure 2.7: Triphenylmethane triisocyanate.

substrates without the brass layer. Isocyanates are very sensitive to moisture and steam and because they are extremely reactive, there is potential for undesirable side reactions with the compounding ingredients in the rubber formulation. Because of their extreme reactivity with moisture, exposure to even moderately humid conditions during the time between application to the metal and the molding process results in loss of much of the adhesive strength of the bond.

Figure 2.8 shows how these adhesive systems respond to temperature. Of these adhesives, cyclized natural rubber is particularly poor and while isocyanates and ebonite give high strength bonds at room temperature, they lose strength rapidly at elevated temperatures.

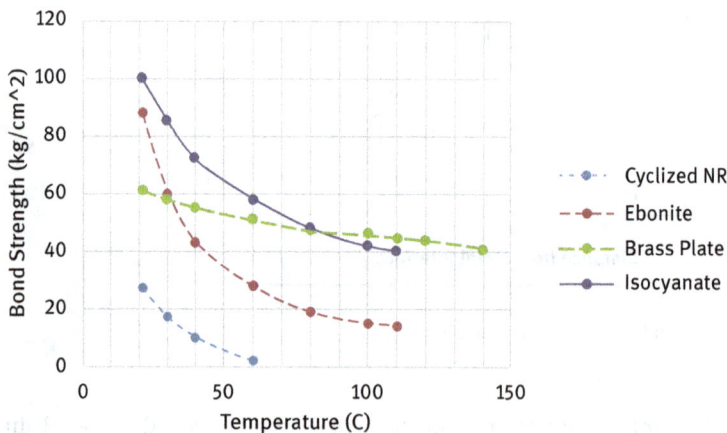

Figure 2.8: Effect of heat on rubber-to-metal bond strength.

During the 1950s, another area of research was based around self-bonding compounds. Patents issued in the late 1950s and early 1960s covered the addition of cobalt salts to the rubber formulation to improve bonding to brass plated steel. The use of resorcinol and a formaldehyde donor (RFL system) was developed in 1935 as a dip for rubber to textile adhesion.

In the 1950s and 1960s, new synthetic polymers were introduced, many of which were not sulfur-curable. Silicone was commercialized around 1945; fluoroelastomers were introduced in the mid to late 1950s and large-scale production of ethylene propylene copolymers began in 1963.

The dearth of versatile and highly satisfactory bonding approaches prompted Lord Corporation to investigate improvements in the rubber to metal bonding process shortly after World War II. The efforts led to the development of general-purpose chemical adhesives containing polymers and cross-bridging agents in solution. The first general purpose adhesive system was introduced in 1956 and was comprised of a primer and a cover-coat that produced rubber tearing bonds over a spectrum of different elastomers, which were commercially available at that time [10]. Migration of crosslinking agents from the adhesive layer into the rubber compound produces a tougher layer adjacent to the metal and increases the tendency to have cohesive failure in the rubber during destructive testing. The primer/cover-coat system gave bonds with better environmental resistance than any other system available at that time and rapidly became the standard for rubber to metal bonding practice. Perhaps more importantly, the new adhesives were broadly compatible with all the important high diene elastomers over a broad range of curing conditions. They paved the way for bonding newer elastomers such as the polychloroprenes and for bonding compounds containing a wider range of compounding ingredients than was previously possible. They also helped to eliminate the restrictions on the design engineer's choice of metal for the substrate. The last four decades have seen the introduction of many new rubber-to-substrate adhesives designed to cover the ever-increasing range of synthetic elastomers currently available for use in dynamic applications. This includes one coat adhesives, adhesives for post-vulcanization bonding, specialty elastomer adhesives for silicones, fluorosilicones, fluoroelastomers, acrylics, and hydrogenated nitrile elastomers, along with the recent introductions of water-based adhesives.

Early adhesives were all solvent based. As more stringent environmental regulations were put in place to restrict solvent emissions, a series of water-based adhesives began to be introduced beginning around 1985. In 2003, new introductions began to eliminate heavy metals, particularly lead and from 2006 to present, the trend was to replace HAP (hazardous air pollutant) solvents with nonHAP solvents.

Today, many companies make adhesives for chemically bonding rubber to metal. The following companies supply general purpose primers and adhesives to the rubber industry:

Company	Tradename
Lord Corporation	Chemlok
Lord Corporation	Chemosil
Dow Chemical	Thixon
Dow Chemical	Megum
Kommerling UK	Cilbond
Axalta	Metalok

References

[1]   Alliger and Sjothun, Vulcanization of Elastomers. Reinhold Publishing Corp, New York, 1964. 117–118.
[2]   German patent 170361.
[3]   Daft, L., inventor, U.S. Patent 1,036,576 to Electro-Chemical Rubber and Manufacturing Company, 1912.
[4]   Daft, L., inventor, U.S. Patent 1,057,333 to Electro-Chemical Rubber and Manufacturing Company, 1913.
[5]   Daft, L., inventor, U.S. Patent 1,057,334 to Electro-Chemical Rubber and Manufacturing Company, 1913.
[6]   Daft, L., inventor, U.S. Patent 1,120,975 to Electro-Chemical Rubber and Manufacturing Company, 1914.
[7]   Rubber Technology and Manufacture, Ed., C. M. Blow, Newnes-Butterworth, London, 1977, 296–298.
[8]   Rubber Technology and Manufacture, Ed., C. M. Blow, Newnes-Butterworth, London, 1977, 296–298.
[9]   German Patent 928,252 Bayer.
[10] Alstadt, D. M., and Coleman, E. W., Jr., U.S. Patent 2,900,292, to Lord Corporation (August 1959).

3 Primers and adhesives

For the organic elastomers, adhesives are either two-part systems comprised of a primer and a topcoat (also referred to as a cover-coat) or they are single coat systems that combine the functions of the primer and the topcoat into a single system. An adhesive is a substance that is able to hold materials together by surface attachment and has measurable film thickness. A primer is a substance that is applied to a substrate to generate good adhesion to the substrate but it also provides a surface that the adhesive topcoat can react with. Silane-based "adhesives" for silicones and specialty elastomers are technically primers and not one-coat adhesives since they have no measurable film thickness.

The adhesive system must be tough enough that it won't fail cohesively through brittle fracture. This toughness is ensured by starting with some type of rubbery polymer dispersed in a solvent or aqueous system for both the primer and the topcoat. The primer provides structural adhesion to the substrate and adhesion to the topcoat. It generally increases the corrosion resistance of a metallic substrate and it is usually a contrasting color to the topcoat so that complete coverage can be seen visually. The topcoat must be capable of crosslinking to the rubber and it must be able to crossbridge to the primer. The topcoat is responsible for the primary bond strength of the joint, but it also has to be able to maintain the strength under tough service conditions while providing resistance to environmental exposure. A single coat system has to provide the functions of primer and topcoat in a single system. It is not created by mixing a topcoat and a primer together as these are generally incompatible. Single coat systems usually exhibit reduced resistance to aggressive environments as compared to two-coat systems.

A crosslinking agent is a molecule having a functionality of two or more that, through chemical reaction, joins two polymeric chains together. These polymer chains are usually the same type and the joints are usually somewhere in the middle of the chains (Figure 3.1). A crossbridging agent is a difunctional molecule that, through chemical reaction, joins polymeric materials together that are on opposite sides of an interface. The interface can be an adhesive–adherent interface or the interface can be between two distinct phases. The polymers on opposite sides of the interface are, more often than not, dissimilar polymers (Figure 3.1).

An elastomer bonded to metal or rigid plastic contains many different complex chemical reactions (Figure 3.2). The primer is thought to be chemisorbed onto the metal or plastic. There will be internal networking or crosslinking in

https://doi.org/10.1515/9783110658996-003

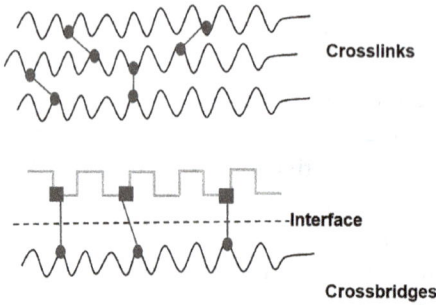

Figure 3.1: Crosslinks versus crossbridges.

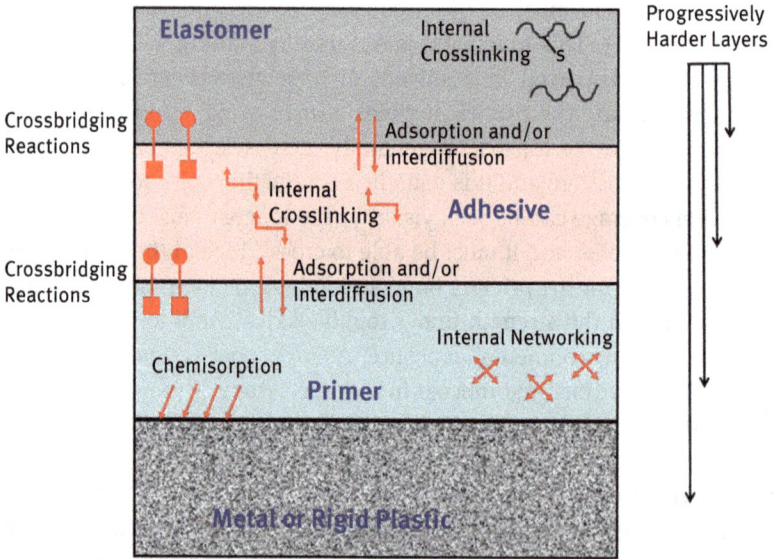

Figure 3.2: Anatomy of a bonded elastomer.

the primer as it reacts and there is some amount of absorption or interdiffusion between the adhesive and the primer. There are chemicals in the primer and adhesive designed to cause a crossbridging reaction. The adhesive itself has internal crosslinking as well as absorption, interdiffusion, and crossbridging between the adhesive and the rubber. Finally, the rubber itself undergoes internal crosslinking. This creates a series of progressively harder layers as we move from soft rubber to rigid substrate [1].

Some of the crossbridging reactions are hypothesized by scientists at LORD Corporation to be due to a Friedel–Crafts reaction involving a chlorinated polymer (Figure 3.3). These reactions produce acid (such as hydrochloric acid or HCl) as a result, so formulations containing halogenated polymers require an acid acceptor to neutralize the acid.

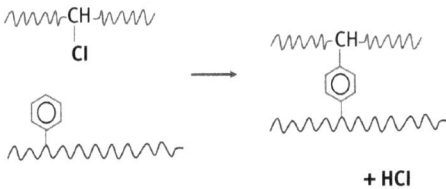

Figure 3.3: Crossbridging hypothesized by a Friedel–Crafts reaction.

The topcoat contains a halogenated polymer (as a film former), inorganic acid stabilizers, cocuratives and crosslinkers, and hot tear strength promotors, all dispersed in a solvent or aqueous carrier. The earliest versions of adhesives had only a minimal number of ingredients. They were made by dissolving chlorinated natural rubber and poly 2,3-dichloro-1,3-butadiene or brominated 2,3-dichloro -1,3-butadiene polymer along with a polyisocyanate in solvent [2]. Shortly thereafter, the polyisocyanate was replaced with a dinitroso aromatic compound as the primary crosslinking and crossbridging agent [3, 4]. The preferred chemical is highly reactive dinitrosobenzene (or DNB) which is still heavily used in adhesives for organic rubber today. As technology moved toward aqueous adhesives, chlorinated polymers continued to be used as the film-formers, but in latex form. Some latexes that were used were 2,3-dichloro-1,3-butadiene [5], chlorinated ethylene/ vinyl acetate copolymer [6], chlorosulfonated polyethylene [7], and other chlorinated polyolefins [8]. The preferred crosslinking additive continued to be DNB although other additives have been disclosed including quinone dioxime (QDO) and polymaleimides [9].

Modern primers typically contain phenolic resins to bond to the metal surface, halogenated polymers to reduce brittleness and to react with the topcoat, metal oxides as acid acceptors, and fillers in either a solvent or aqueous carrier [10]. One of the earliest primers used with modern adhesive topcoats contained chlorinated natural rubber, an epoxidized phenolic resin, and an epoxy resin curing agent all dissolved in solvent [11]. While chlorinated natural rubber still finds use in primers, more heat and oil resistant primers can be made using chlorinated polyolefins [8] or neoprene (polychloroprene) as the film-forming polymer [12]. The primer may also contain such things as phenoxy resins and/ or silanes [13]. Even the variety of chemicals that can be added to primers and topcoats as acid stabilizers is lengthy. From US patent #5,496,884, we read:

> The acid-scavenging compound is preferably a metal oxide or a lead-containing compound. The metal oxide of the present invention can be any known metal oxide such as the oxides of zinc, cadmium, magnesium, lead, and zirconium; litharge; red lead; zirconium salts; and combinations thereof. Various lead-containing compounds may also be utilized as an acid-scavenging compound in lieu of, or in addition to, the metal oxide. Examples of such lead-containing compounds include lead salts such as polybasic lead salts of phosphorous acid and saturated and unsaturated organic dicarboxylic acids and acid anhydrides. Specific examples of lead salts include dibasic lead phthalate, monohydrous tribasic lead maleate, tetrabasic lead fumarate, dibasic lead phosphite, and combinations thereof. Other examples of lead-containing compounds include basic lead carbonate, lead oxide and lead dioxide [9].

One-coat systems contain a plethora of ingredients designed to act as both the primer and adhesive in one package. One example in the patent [14] requires the following components at a minimum:
a) a Diels–Alder adduct of a perhalogenated cyclic conjugated diene and an olefinically unsaturated dienophile having a vinyl content in excess of 50 percent
b) a phenolic resin
c) an aromatic hydroxy compound
d) a formaldehyde donor
e) a heat-activated unsaturated elastomer crosslinker
f) a vulcanizing agent
g) a metal oxide

Many of the active ingredients in both primers and adhesive topcoats are not soluble but rather are in a suspension and they settle out over time. For this reason, it is important to adequately stir both primers and adhesives well before applying them to a substrate.

It should also be noted that chemicals from the adhesive can migrate into the rubber and vice versa. A conical bond specimen (see Chapter 5) was bonded using clear polybutadiene. After bonding, it can be seen that the dark staining DNB crosslinker in the adhesive has migrated several millimeters into the clear polybutadiene compound evidenced by the orange halo in the rubber near the method C conical metal inserts (Figure 3.4).

This means that the adhesive can alter the properties of the rubber in the thin film that is near the adhesive interface. Likewise, ingredients from the rubber can migrate into the adhesive film, sometimes causing antagonistic effects (more about that in Chapter 8).

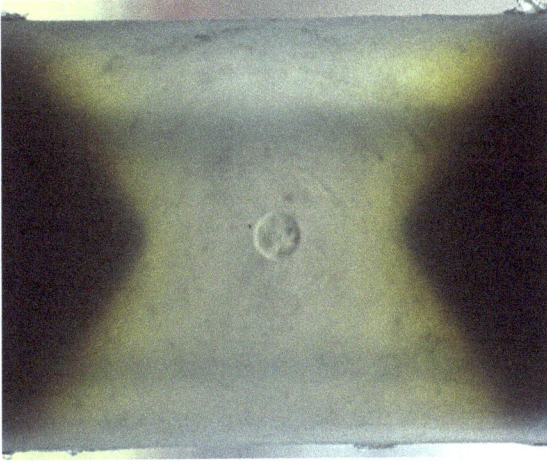

Figure 3.4: DNB migration from the bond line into the clear cured elastomer.

Although most bonds are formed during the process of vulcanization, some rubber parts are bonded to a substrate after curing. This is referred to as post-vulcanization bonding or PV bonding. Only a few adhesives are recommended for PV bonding, so it is important to choose the adhesive system very carefully. PV bonds are never as robust as those formed during the vulcanization process.

Silicone rubber and some other specialty hydrocarbon elastomers may be bonded using silane-based chemistry. Technically, silane is a primer and not an adhesive. There is no measurable film thickness since it is only a few atoms thick after application. The silane (Figure 3.5) has an organophilic end R1 and inorganophilic groups R that are hydrolysable. The inorganophilic ends bond to the hydroxyl (–OH) groups on the surface of the metal after hydrolysis. The organophilic end (R1) is functional and is selected for reactivity with the rubber. The hydrolysable groups are most commonly either methoxy or ethoxy.

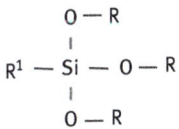

$$
\begin{array}{c}
O - R \\
| \\
R^1 - Si - O - R \\
| \\
O - R
\end{array}
$$

Figure 3.5: Structure of a typical silane.

The bonding reaction occurs in three steps (Figure 3.6). Silanes are moisture sensitive. During the hydrolysis step, the silane reacts with three water molecules to split off methyl or ethyl alcohol (depending on whether it is a methoxy or ethoxy silane). When a metal part is dipped into the primer, the –OH groups on the

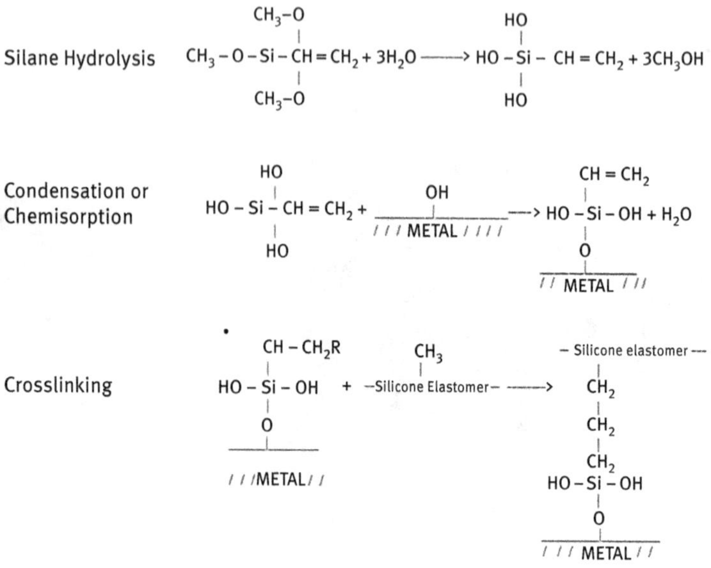

Silane Hydrolysis

$$
\underset{\underset{CH_3-O}{|}}{\overset{\overset{CH_3-O}{|}}{CH_3-O-Si-CH=CH_2}} + 3H_2O \longrightarrow \underset{\underset{HO}{|}}{\overset{\overset{HO}{|}}{HO-Si-CH=CH_2}} + 3CH_3OH
$$

Condensation or Chemisorption

$$
\underset{\underset{HO}{|}}{\overset{\overset{HO}{|}}{HO-Si-CH=CH_2}} + \underset{///\ \text{METAL}\ ///}{\overline{\underset{|}{OH}}} \longrightarrow \underset{\underset{\overline{//\ \text{METAL}\ //}}{\overset{|}{O}}}{\overset{\overset{CH=CH_2}{|}}{HO-Si-OH}} + H_2O
$$

Crosslinking

$$
\underset{\underset{\overline{///\text{METAL}//}}{\overset{|}{O}}}{\overset{\overset{CH-CH_2R}{|}}{HO-Si-OH}} + \overset{\overset{CH_3}{|}}{-\text{Silicone Elastomer}-} \longrightarrow
$$

$$
\begin{array}{c} -\ \text{Silicone elastomer}\ --- \\ | \\ CH_2 \\ | \\ CH_2 \\ | \\ CH_2 \\ HO-Si-OH \\ | \\ O \\ \overline{///\ \text{METAL}\ //} \end{array}
$$

Figure 3.6: Reaction process for bonding with a silane primer.

silane can condense with –OH groups on the metal to form an attachment. During the crosslinking reaction, the rubber reacts with the functional group on the silane to bond the rubber to the metal.

For silicone and for some peroxide-curable elastomers, the functional group is often vinyl. Certain other specialty polymers like some of the fluorocarbon elastomers may be bonded with a silane primer but the functional groups might be amino or glycidoxy instead. Even these are not as simple as coating the metal part with a silane. The silane must be diluted, prehydrolyzed, and stabilized. Often, the primers contain a mixture of different silanes.

This commentary barely scratches the surface of commercially available adhesive and primer formulations. The reason there are so many adhesives on the market is because no one primer or adhesive will begin to bond all the types of rubber to all the substrates that are used in industry. For this reason, it is best to select an adhesive system in consultation with the adhesive supplier.

References

[1]   Lee, L. H., Adhesion Science and technology, Plenum Press, New York, 1975, 265.
[2]   Alstadt, D. M., and Coleman, E. W., Jr., U.S. Patent 2,900,292, to Lord Corporation (August 1959).
[3]   Alstadt, D. M., and Coleman, E. W., Jr., U.S. Patent 3,258,388, to Lord Corporation (June 1966).
[4]   Alstadt, D. M., and Coleman, E. W., Jr., U.S. Patent 3,258,389, to Lord Corporation (September 1959).
[5]   Sadowski, J. S., U.S. Patent 4,483962, to Lord Corporation (November 1984).
[6]   Gervase, N. J., and Manino, L. G., U.S. Patent 5,367,010, to Lord Corporation (November 1994).
[7]   Mowrey, D. H., U.S. Patent 5,281,638, to Lord Corporation (January 1994).
[8]   Ozaya, H., et al., U.S. Patent 5,385,979, to Lord Corporation (January 1995).
[9]   Weih, M. A., et al., U.S. Patent 5,496,884, to Lord Corporation (March 1996).
[10]   Kucera, H. W., U.S. Patent 6,476,119, to Lord Corporation (November 2002).
[11]   De Crease, W. M., U.S. Patent 3,099,632, to Lord Corporation (July 1963).
[12]   Mowrey, D. H., et al, U.S. Patent 5,093,203, to Lord Corporation (March 1992).
[13]   Gervase, N. J., U.S. Patent 4,308,071, to Lord Corporation (December 1981).
[14]   Warren, P. A., et al., U.S. Patent 5,128,403, to Lord Corporation (July 1992).

4 Rubber chemistry

The chemistry of the rubber reacts with or interacts with the chemistry of the adhesive, whether synergistically or antagonistically. It is, therefore, useful to understand the basic concepts of rubber chemistry. There are thousands of different ingredients that find use in rubber formulations and the formulations can be both complex and diverse. Elastomers (rubbery polymers) can be divided into two classes: inorganic elastomers and organic elastomers. Polymers that comprise a carbon backbone are termed organic and include most of the elastomers that are in use with the exception of silicones. Silicones have an alternating silicon–oxygen backbone.

An elastomer is a polymeric material that, at room temperature, can be stretched repeatedly to at least twice its length, and, upon release of the stress, returns approximately to its original length. Polymers are high-molecular-weight compounds made up of low-molecular-weight building blocks called monomers. There may be 1,000–20,000 repeating units of the monomer in a typical polymer chain. These monomers are linked end to end and are free to rotate or bend at any of the carbon–carbon connecting points. In the monomeric form, butadiene is a gas but when several thousands of these monomeric units are linked together, they form polybutadiene which is a rubbery solid.

Rubber formulations are recipes with multiple ingredients selected to meet a balance of properties in the final compound. The basic ingredients in an elastomer formulation are:
- Polymer/elastomer
- Cure system
- Filler/reinforcement
- Plasticizers and process aids
- Antidegradants
- Miscellaneous additives

Miscellaneous additives include colors, abrasives, conductive agents, blowing agents, lubricants, fungicides, flame resistors, and acid acceptors.

There are three goals for the compounds: functionality of the final product, processability of the compound, and cost. Depending on the product and market, cost may not matter or cost may be very important. Every ingredient in the list above is varied for purposes of function first, and for cost secondarily, except for fillers and plasticizers. These may be used at high levels to dilute the polymer

https://doi.org/10.1515/9783110658996-004

and lower the final compound cost, when performance requirements are minimal. Producing a floor mat and making an aircraft bearing are enormously different challenges.

Formulations are written in phr, which stands for parts per hundred rubber. This is because most ingredients react with the polymer (like curatives and antidegradants) or they modify the polymer (like plasticizers and reinforcing agents). phr makes it easy to compare different formulations and see the other ingredient levels relative to that of the polymer. By convention, formulations start with 100 phr total of the polymer or mixture of polymers in the formulation. A simple recipe for a general-purpose rubber might look like what is given in Table 4.1.

Table 4.1: Typical formulation for a general-purpose rubber.

Ingredient	phr	Function
Styrene butadiene rubber	100	Basic material
Carbon black	50	Reinforcement
Oil	15	Softener, process aid
Zinc oxide	5	Aids vulcanization
Stearic acid	1	Processing, curing
Antioxidant	2	Improves aging
Antiozonant	2	Improves aging
Waxes, esters	2	Process aids
Sulfur	2	Makes cross-links
Accelerators	3	Speeds up cross-link formation

It is useful to think of polymers as having spaghetti-like structure where the strands are in constant random motion. In an uncross-linked state, these polymer chains are free to flow past one another, and like any liquid, they readily take on new shapes during the mixing and molding operations. The everyday example of uncross-linked rubbery polymer is chewing gum.

Vulcanization, otherwise known as curing, is the conversion of polymer molecules into a reasonably permanent three-dimensional network by the formation of cross-links that change the rubber from a very viscous liquid into an elastic solid. Vulcanization forms chemical bonds between the polymer chains that prevent them from sliding past one another.

Two characteristics of the polymer are important to the bonding of elastomers: unsaturation and polarity.

Unsaturation (the presence of double bonds) is important because it provides reactive sites through which vulcanization or adhesive bond formation can occur (Figure 4.1). Unsaturation in the backbone of a polymer provides sites that are susceptible to attack from oxygen, ozone, oxidizing chemicals, and radiation, but it also allows a polymer to be more easily bonded.

Unsaturation

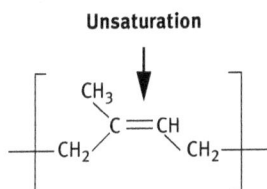

Figure 4.1: Unsaturation (double bonds) in each monomeric unit of polyisoprene (the polymer that makes up natural rubber).

Polymers with no unsaturation cannot be sulfur-cured. Some polymers like ethylene–propylene–diene rubber (EPDM) have a small amount of pendant unsaturation, which allows it to be sulfur-cured (Figure 4.2). The amount of unsaturation in EPDM generally runs between 4% and 10% and it is pendent to the polymer backbone.

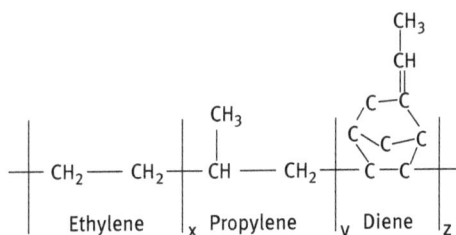

Figure 4.2: Pendant unsaturation (double bonds) in EPDM rubber.

Molecules are commonly classified by polarity, which is based on the distribution of electrical charges in the molecule. Nonpolar substances either have comparatively electrically neutral bonds (such as between carbon and hydrogen) or have a symmetrical structure (e.g., carbon tetrachloride or CCl_4). A polar molecule is one in which there is asymmetric separation of positive and negative charges, and there are of course varying degrees of polarity. Water or H_2O is a nonlinear molecule with partial charge separation (oxygen is more electronegative than hydrogen) and is an example of a polar material.

In the hydrocarbon-based polymers, the presence of atoms other than carbon and hydrogen will impart varying degrees of polarity. Polyisoprene and polybutadiene are composed of only hydrogen (H) and carbon (C) atoms and are both nonpolar. EPDM (Figure 4.2) is also comprised only of carbon and hydrogen so it is nonpolar as well. Nitrile rubber, a copolymer of butadiene and acrylonitrile, is a polar polymer (Figure 4.3) because of the nitrogen (N) in the acrylonitrile segments. In general, polar polymers are easier to bond than nonpolar polymers.

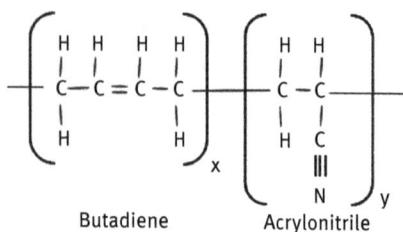

Butadiene   Acrylonitrile

Figure 4.3: Nitrile rubber is polar because of the nitrogen in the acrylonitrile.

Table 4.2 shows the unsaturation level and the relative polarity of some common families of elastomers identified first by the ASTM abbreviation and secondarily by the chemical name of the polymer.

Table 4.2: Polarity and unsaturation for elastomers.

Polymer	Chemical name	Polarity	Unsaturation
FKM	Fluorocarbon elastomers	Very high	None
ACM	Acrylic	Very high	Low
AEM	Ethylene acrylic	Medium-high	Low
CO/ECO	Epichlorohydrin	High	Low
NBR	Nitrile–butadiene	High	High
HNBR	Hydrogenated NBR	High	Low/very low
CM	Chlorinated polyethylene	Moderate	Low
CR	Neoprene	Moderate	High
IIR	Butyl	None	Low
SBR	Styrene–butadiene	None to very low	High
BR	Polybutadiene	Nonpolar	High
NR	Natural rubber	Nonpolar	High
IR	Polyisoprene	Nonpolar	High
EPDM	Ethylene–propylene	Nonpolar	Low
VMQ/PVMQ	Silicone	Moderate	Very low
FVMQ	Fluorosilicone	Very high	Very low

Ease of bonding is largely determined by the combination of the polarity of the polymer and the level of unsaturation. The general-purpose elastomers are often assigned a bondability index ranging from the easiest to bond to the most difficult (Figure 4.4).

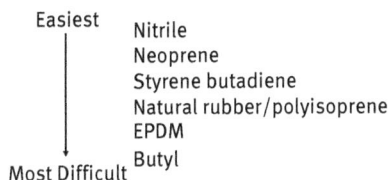

Easiest
Nitrile
Neoprene
Styrene butadiene
Natural rubber/polyisoprene
EPDM
Most Difficult Butyl

Figure 4.4: Bondability index for general-purpose elastomers.

We can subdivide fillers into three categories: reinforcing, semireinforcing, and extending. To be reinforcing, fillers must have primary particle sizes in the range of 10–100 nm (0.01–0.10 µm). Semireinforcing fillers range from 0.1 to 2.0 µm and anything larger functions merely as an extending filler.

The most commonly used reinforcing fillers in rubber are silica and carbon black. These fillers confer strength to the compounds. Semireinforcing fillers include hard clay, talc, and calcium silicate. Extending fillers cover a wide range of materials including aluminum hydroxide, aluminum silicate, barium sulfate, ground limestone, oyster shell or chalk, ground sand, soft clays, graphite, mica powder, zinc or magnesium oxide, and flock.

Most compounds contain a reinforcing filler but the surface chemistry between carbon black and silica is quite different. Silica has mostly OH groups on the surface that does not interact with most organic polymers without some type of surface treatment. Carbon black, on the other hand, has a multitude of organic groups on the surface that promotes interaction with the organic polymer (Figure 4.5).

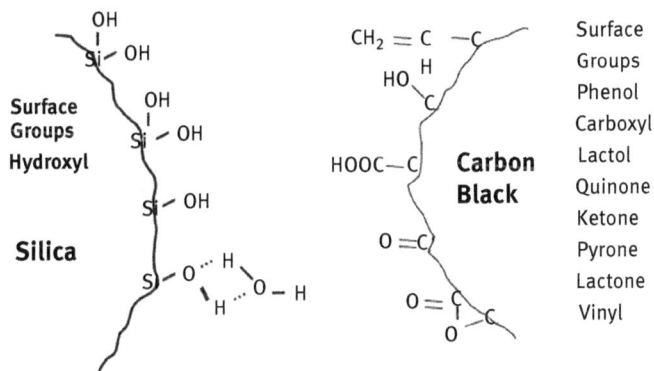

Figure 4.5: Surface chemistry differences between carbon black and silica.

The surface chemistry of silica is easily modified using silanes. Methoxy or ethoxy groups on silanes react with the hydroxyl (OH) groups on the silica surface (Figure 4.6). Silanes generally have one and sometimes two organofunctional groups (abbreviated as "R"). The organofunctional group is chosen based on the chemistry of the polymer and cure system. For commonly available silanes, R is chosen from the following reactive groups:

- Amino
- Mercapto
- Chloro
- Vinyl
- Methacryl
- Epoxy
- Isocyanato
- Alkyl (octyl)
- 3-Thiocyanato
- Sulfur

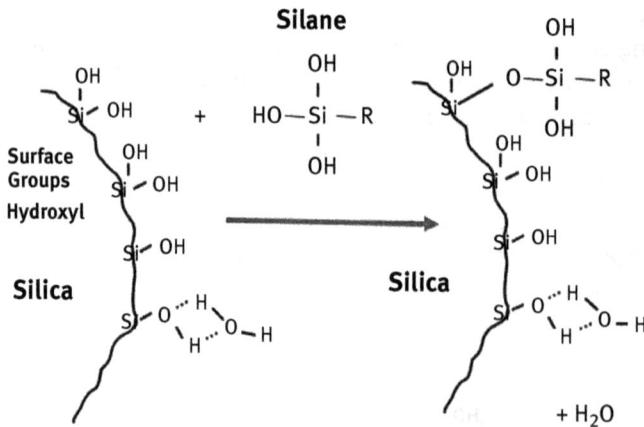

Figure 4.6: The chemistry of the silica surface can be modified using silanes.

With the exception of strain crystallizing polymers like polyisoprene and polychloroprene, the other elastomers require reinforcing fillers for strength. For example, the strength of EPDM can be increased by an order of magnitude with reinforcing carbon black.

EPDM elastomer	No black	50 phr black
Tensile strength (MPa)	1.4	18.6
Elongation (%)	275	325

These reinforcing fillers also aid in bond formation with the adhesive.

Vulcanization (also known as cross-linking or curing) is a nonreversible thermosetting reaction. The cross-linking or curing reaction occurs when the rubber compounds are exposed to heat for a given amount of time. Table 4.3 shows a list of some of the materials that can be used to cross-link elastomers.

Table 4.3: Cross-linking agents.

Sulfur
Sulfur donors
Metal oxides
Thioureas
Resins
Quinones
Peroxides
Polyfunctional amines
Bisphenols
Isocyanates
Radiation

Sulfur and peroxide are perhaps the two most common vulcanizing agents. Entire books have been written on vulcanization and vulcanizing agents, and still, the details of the chemistry are not always agreed upon. Each elastomer has its own peculiarities in its response to curing. Curing may enhance or degrade a desirable property, and formulating to attain one optimum characteristic may seriously reduce some other desired property.

A rheometer is a trace that tracks vulcanization by change in viscosity and modulus. Vulcanization occurs in three stages: (1) an induction period during which the compound can flow or take a shape; (2) a curing or cross-linking stage; and in some cases (3) a reversion or over-cure stage (Figure 4.7).

The induction period represents the time interval at curing temperatures when no measurable cross-linking can be observed. Following the induction period, cross-linking occurs at a rate that is dependent on the formulation and on the temperature. When cross-linking has proceeded to full cure, subsequent

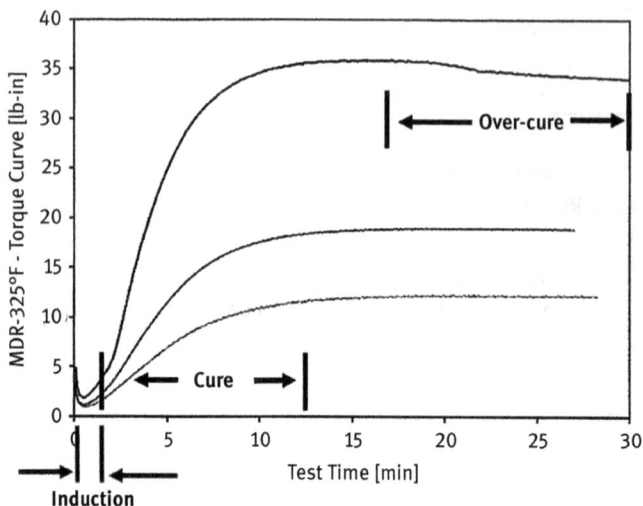

Figure 4.7: Rheometer traces showing three stages of the cure reaction.

heating may produce an over-cure, which may be evidenced by continued stiffening or by reversion (softening). Reversion is evidenced by a reduction in modulus and tensile strength and an increase in elongation. As the cure temperature increases, cure times become shorter, but reversion becomes more pronounced also.

Sulfur is combined in the vulcanization network of diene rubbers (NR, BR, SBR, NBR, CR, EPDM, etc.) in many ways. In cross-links, sulfur may be present as monosulfide, disulfide, or polysulfide chains. High sulfur levels with low accelerator levels produce primarily polysulfidic cross-links while low sulfur levels and high accelerator levels produce primarily mono- and disulfidic links. Peroxide cure systems produce direct carbon-to-carbon cross-links (Figure 4.8).

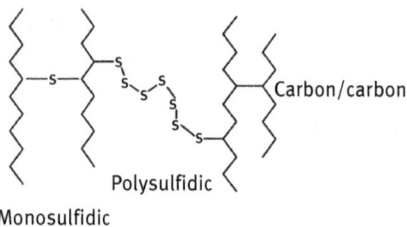

Figure 4.8: Types of cross-links.

Although the diene elastomers can be cured with sulfur alone, sulfur accelerators are generally used to speed up the curing reaction. There are a number of families of sulfur accelerators and each family contains multiple different chemicals (Table 4.4). The first four are most commonly used and they range in order from slower to faster in reaction rate.

Table 4.4: Families of sulfur accelerators.

Sulfenamides
Benzothiazoles
Thiurams
Dithiocarbamates
Triazines
Guanidines
Aldehyde-amines
Dithiophosphates

For the rheometer graphs in Figure 4.9, the following accelerators were used along with 2 phr sulfur in a natural rubber gum stock containing 5 phr zinc oxide and 1 phr stearic acid:

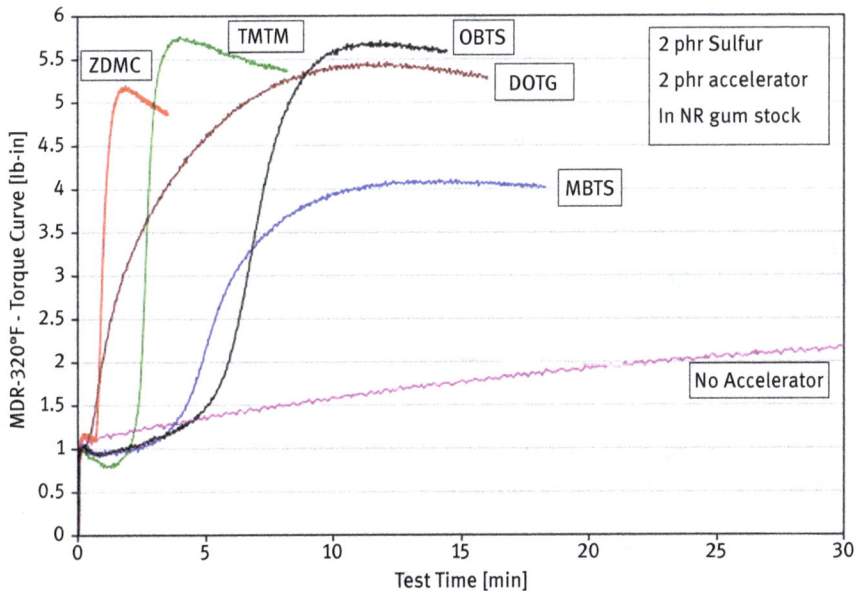

Figure 4.9: Comparison of five accelerator types.

OBTS	Sulfenamide (*n*-oxidiethylene-2-benzothiazole sulfenamide)
MBTS	Thiazole (2-mercaptobenzothiazole disulfide)
TMTM	Thiuram (tetramethylthiuram monosulfide)
ZDMC	Dithiocarbamate (zinc dimethyldithiocarbamate)
DOTG	Guanidine (diorthotolyl guanidine)

Peroxide cross-linking involves the generation of free radicals from the dissociation of a peroxide with heat. Organic peroxides are characterized by at least one oxygen-to-oxygen bond (Figure 4.10).

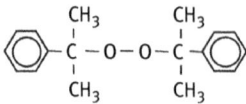

Figure 4.10: Dicumyl peroxide.

The peroxide reaction sequence can be broken into three reactions. The first is to split the organic peroxide into peroxide radicals, the decomposition step. This step is the rate-determining step in the reaction. The second step is forming a radical on the polymer radical formation through hydrogen abstraction from the polymer or by addition to the polymer chain. The third and final step is polymer radical recombination, which is when cross-linking occurs. The efficiency of the cross-linking reaction depends on the type of peroxide used and the type of polymer radical formed. Polymer polarity, the number of reactive groups, and the compounding ingredients and fillers used have some impact on cure speed and efficiency.

The following elastomers are cross-linkable with organic peroxides:

NR	Natural rubber
IR	Polyisoprene
BR	Polybutadiene
CR	Polychloroprene
SBR	Styrene butadiene
NBR	Nitrile
HNBR	Hydrogenated nitrile
Q	Silicone/fluorosilicone
EPM/EPDM	Ethylene propylene
PE/POE	Polyethylene/polyolefin
CM	Chlorinated polyethylene
CSM	Chlorosulfonated polyethylene
EVA	Ethylene vinyl acetate
EAM	Ethylene acrylic
FKM	Certain grades of fluoroelastomer
TFEP	Tetrafluoroethylene propylene

Coagents increase the cross-link density and may allow for less peroxide to be used. Some coagents speed up the early stages of cross-linking. Figure 4.11 shows that the addition of phenylene bismaleimide (HVA-2) to 4 phr of dicumyl peroxide significantly increases the cross-link density. Coagents alone show little or no cross-linking without a peroxide present. They are reactive materials with a functionality of at least two and they become a part of the cross-link network. Some of the commonly used coagents have functional groups that include vinyl, allyl, acrylate or methacrylate, and bismaleimide. However, not every coagent works with every polymer.

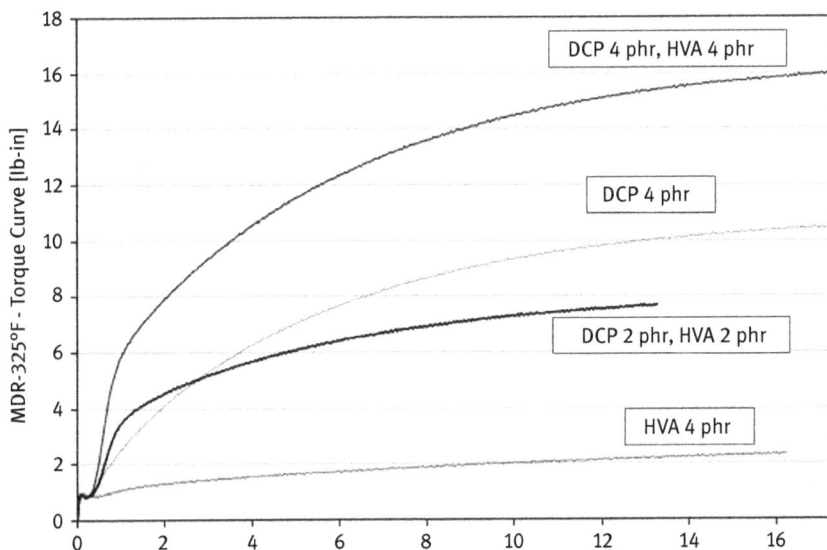

Figure 4.11: Phenylene bismaleimide coagent in a natural rubber gumstock.

Plasticizers are used for many purposes in rubber. They can decrease the elastomer cost by using high levels of less expensive carbon black and oil. In small quantities, they improve the flow of rubber during mixing, processing, and molding. They can improve the filler dispersion in the compound. They can be used to influence various physical properties of the compound. Plasticizers soften a rubber compound, and certain plasticizers improve low-temperature flexibility or render the rubber more flame retardant. Table 4.5 shows a list of some of the types of plasticizers used in rubber.

Regardless of the reasons for using plasticizers, compatibility is of major concern. Every polymer has its own level of compatibility with various oils,

Table 4.5: Plasticizers in rubber.

Petroleum oils
Naphthenic
Paraffinic
Aromatic
Esters and polyesters
Phthalates
Adipates
Glutarates
Sebacates
Azelates
Oleates
Trimellitates
Ethers
Paraffin waxes
Coumarone and indene resins
Polymeric plasticizers
Petroleum distillation residues including bitumen and pitch
Factice (vulcanized vegetable oil)
Fatty acids and metallic soaps
Pine tar and rosin
Chlorinated hydrocarbons
Vegetable oils
High styrene resins
Liquid polymers

which means that some oils will be acceptable at very high levels, others at moderate levels, and some only at very low levels. If an oil is used at a level beyond its solubility in the polymer, it will bleed out, causing poor physical properties, a sticky surface, and poor adhesion. Compatibility is greatest when the solubility parameters of the plasticizer and polymer are similar. As an example, Figure 4.12 is a picture of a cured piece of EPDM rubber that was formulated with soybean oil.

Antidegradants are used in elastomers to protect against heat, oxygen, ozone, fatigue, light including ultraviolet, hydrolysis, crazing, frosting, and metal poisoning. In a diene elastomer, it only requires 1–2% oxygen for degradation to occur and the rate of oxidation doubles for each 10 °C temperature rise. Oxidation can cause additional cross-linking (which results in hardening and embrittlement), chain scission (which results in softening), and chemical alteration of the polymer chain. There are a number of classes of antidegradants (Table 4.6) with multiple chemical compounds in each class.

Figure 4.12: Cured EPDM with soybean oil bleeding to the surface.

Table 4.6: Classes of antidegradants for rubber.

Phenylenediamines
Dihydroquinolines
Naphthylamines
Diphenylamines
Benzimidazoles
Bisphenols
Monophenols
Phosphites
Petroleum waxes
Other miscellaneous chemicals

Silicone is a general term for organopolysiloxanes, the most common of which is polydimethylsiloxane, sometimes abbreviated as PDMS. The backbone of silicone is inorganic, an alternating silicon–oxygen chain (-Si-O-). The side groups are organic, which makes silicone a hybrid inorganic–organic polymeric material.

Silicone gums are classified based on their structure. Q refers to the Si–O linkage, and in particular, the silicon–oxygen backbone. Commercially available HCR (high consistency rubber) silicone elastomers can be divided into three broad classifications:

(1) Polydimethylsiloxanes (MQ or VMQ)
(2) Copolymers of polydimethylsiloxane with diphenyl or phenylmethyl siloxane (PMQ or PVMQ)
(3) Poly-3,3,3-trifluoropropylmethyl siloxane (better known as fluorosilicone) (FMQ or FVMQ)

Side groups are abbreviated as P, V, M, and F. M refers to methyl, so MQ is a straight dimethyl gum. "V" stands for vinyl, which is added in small quantities as a cure site. "P" indicates phenyl and "F" refers to a fluorinated silicone. The simplest silicone gum is MQ or (di)methyl silicone. A small amount of vinyl (generally about 0.1 mole percent) can be added as a cure site, making a VMQ or vinyl methyl silicone.

Reinforcing fillers must be used in silicone elastomers since the tensile strength of an unreinforced cured silicone gum is extremely low, on the order of 0.15 MPa. The most commonly used reinforcing fillers are the pyrogenic or fume-process silicas. Even with reinforcing silica, silicone rubber usually has poor tensile strength relative to organic elastomers, peaking out between 10 and 12 MPa.

Traditional antioxidants are not used in silicone elastomers, but heat stabilizers are generally incorporated. One of the best and also cheapest is red iron oxide or ferric oxide used at 1–2 phr. Ferric oxide also has an advantage in that it gives minimal inhibition of peroxide cures. Other important stabilizers are barium zirconate (4 phr) and ceric oxide, which are especially useful in white or colored stocks, and thermal black which is used at 0.1–2.0 phr. Numerous other metal derivatives have been evaluated, especially transition metal and rare earth derivatives, but they seldom exceed the effectiveness of the more common and less expensive ferric oxide.

HCR silicones are generally cured using organic peroxides. A number of different peroxides are available and the properties of the cured elastomer will be influenced by the peroxide choice. Organic peroxides are compounds that contain one or more oxygen–oxygen bonds (R-O-O-R) in the molecule. The structure of the peroxide will affect its thermal stability, its chemical stability, and the energy of the free radicals generated. Although there are seven classes of organic peroxides, two classes of peroxides are commonly used for cross-linking silicone rubber. Diacyl peroxides (also sometimes referred to as diaryl peroxides) are based on carboxylic acids and are nonspecific, producing high-energy radicals capable of hydrogen abstraction. These can be used to cure dimethyl silicones containing no vinyl. Dibenzoyl peroxide is a common example of a diacyl peroxide (Figure 4.13).

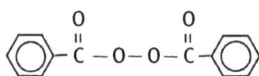

Figure 4.13: Dibenzoyl peroxide.

Diacyl peroxides typically cure at lower temperatures and the decomposition products are carboxylic acids. Since silicone is sensitive to acids, an extended postcure is required to drive off the acid residue from the compounds. Failure to remove the acidic decomposition products will result in depolymerization of the silicone compound over time during service.

Even though they will cure dimethyl silicones, the diacyl peroxide radicals react even more quickly with vinyl so they are effective in vinyl methyl silicones as well. Diacyl peroxides cannot be used with silicone compounds containing significant quantities of carbon black reinforcement because the black interferes with the curing reaction. Carbon black, an important reinforcing filler for organic elastomers, does not reinforce silicone polymers as well as silica and in quantity, it detracts from thermal stability, but it is used as a colorant or when making conductive silicone rubber.

Dialkyl peroxides produce lower energy radicals, which are more stable and do not readily abstract hydrogen. They are essentially vinyl specific, curing only through vinyl groups. Today, most silicone gums contain vinyl-cure sites, generally about 0.1 mole percent. The organic peroxide 2,5-bis(tert-butylperoxy) 2,5-dimethyl hexane is a commonly used example of a dialkyl peroxide (Figure 4.14).

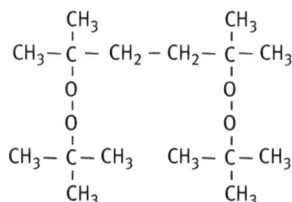

Figure 4.14: Dialkyl peroxide 2,5-bis(tert-butylperoxy) 2,5-dimethyl hexane.

The decomposition products of the alkyl peroxides are alcohols instead of acids. Alcohols are more easily removed with a short postcure and, unlike carboxylic acids, if they are not removed, they do not cause chain scission, which depolymerizes the rubber over time during service.

Another important cure mechanism involves hydrosilylation with platinum catalysts. This mechanism is particularly important for liquid silicone rubbers (LSR). LSR silicones have a major production advantage in that their cure times can be extremely short, seconds rather than minutes. The addition cure reaction involves the addition of polyfunctional silicon hydride (Si-H) to unsaturated

groups (usually vinyl groups) in the polysiloxane chain. This reaction is catalyzed using a platinum complex such as chloroplatinic acid. Elastomers cured
with platinum-catalyzed addition cures show exceptional toughness and tensile
strength, a tight surface cure, and nonyellowing translucency. The platinum-
cured silicones are usually postcured to complete the cross-linking process. One
of the downsides to this type of cure system is it results in a much shorter shelf
life of the catalyzed compound than when peroxides are used. These systems are
usually supplied in two parts. By convention, part A contains the vinyl silicone
polymer and a small amount of platinum catalyst, typically 5–10 ppm. The B side
contains the hydride functional silicone polymer. Platinum cures are quite sensitive to cure inhibition from contamination with trace quantities of certain chemicals, such as sulfur or amines, commonly used in organic elastomers.

4.1 Self-bonding compounds

One question that often comes up is whether elastomer compounds can be
made to be self-bonding. The answer is, "Certainly." The tire industry routinely
uses self-bonding compounds to bond to brass-plated steel, but only with high-
diene elastomers with a reasonably high sulfur cure [1, 2]. These compounds
usually include resorcinol resins, cobalt compounds (like cobalt naphthenate),
and a methylene donor (like hexamethoxymethylmelamine, also known as
HMMM). But what about other elastomers and other substrates? The answer is
a qualified yes. If the main purpose of the bond is to hold the rubber in place
through storage and installation, self-bonding compounds may be useful, especially if the rubber is placed into compression during service. There are two
main drawbacks to making a compound self-bonding. The first problem is that
the bond generally lacks environmental resistance and it may fail during service in aggressive environments. The second problem is that we are modifying
the entire elastomer formulation in order to affect the chemistry at the interface.
This generally has an impact on the properties of the entire compound when
using an adhesive affects the chemistry only at the substrate interface and in
the rubber very close to the interface. When you try to change one property,
bondability for example, you change other properties often in ways that are not
favorable (Figure 4.15). In the case of the nitrile, the hardness was significantly
increased while the elongation was significantly decreased. For the fluorocarbon, the hardness decreased and the elongation increased, which indicates
some sort of antagonism.

For sulfur-cured elastomers, the use of maleic-anhydride-adducted polybutadiene has been proposed, typically at 10 phr [3, 4]. Other technology that has

Ingredient	NBR Control -1	NBR -2	FKM Control -3	FKM -4
Nipol 1041 (NBR)	100.00	100.00		
Viton GBL-200S (FKM)	-----	-----	100.00	100.00
STEARIC ACID	1.00	1.00	-----	-----
ZINC OXIDE	5.00	5.00	3.00	3.00
Flectol TMQ	2.00	2.00	-----	-----
VPA #2 Process Aid	-----	-----	1.00	1.00
N-990 Black	-----	-----	30.00	30.00
N550 Black	50.00	50.00	-----	-----
Varox DBPH-50 (peroxide)	-----	-----	1.50	1.50
TAIC	-----	-----	2.50	-----
Dicumyl peroxide 40%	5.00	5.00	-----	-----
Self-Bonding agent xx1	-----	10.00	-----	-----
Self-Bonding agent xx6	-----	-----	-----	2.50
Rheometer @ 350°F				
Low torque	2.03	1.62	0.50	0.57
High torque	**17.89**	**25.61**	**23.96**	**20.50**
Tc90 (minutes)	3.93	3.69	1.49	1.53
PHYSICAL PROPERTIES				
Hardness (Shore A)	73	**81**	75	**73**
Tensile (PSI)	3467	3462	2735	**2303**
Elongation (%)	307	**215**	313	**376**
100% modulus	825	1609	596	**454**

Figure 4.15: Addition of self-bonding agents to a nitrile and a fluorocarbon compound.

been proposed for organic elastomers are that of metallic coagents, particularly zinc diacrylate (ZDA) and zinc dimethacrylate (ZDMA) [5, 6]. These are generally used in peroxide-cured compounds because the free radicals are able to react the acrylate groups.

For silicones, there are many patents, both active and expired, covering rubber to substrate adhesion. Most commonly, they rely on the addition of fumerates or maleates [7], or silanes, particularly epoxysilane in combination with other chemicals [8–10].

References

[1] A Review of Bonding Agents as Adhesion Promotors in Rubber to Metal and Rubber to Textile Applications. Seibert, R., Paper 52, 144th Meeting of the Rubber Division ACS. October, 1993.
[2] A Review of In-Compounding Bonding Agents. Weaver, E. J., edited by D. Walker, published in Rubber and Plastics News, 07/10/1978.
[3] Using Polybutadiene Derived Resins for Improved Elastomer Bonding. Drake, R., Labriola, J., and Sessions, H., Paper 55, 138th Meeting of the Rubber Division ACS. October, 1990.
[4] Adhesion Promotion using Maleated Polybutadiene Resins in Rubber Compounds. Drake, R. and Labriola, J., Paper 48, 139th Meeting of the Rubber Division ACS. May, 1991.

[5]   Steiber, J. F., Hong, S. W., and Seibert, R. F., U.S. Patent 5,217,807A, to Chemtura Corporation (June 1993).

[6]   Metallic Coagents for Rubber-to-Metal Adhesion. Costin, R. and Nagel, W., Published by Cray Valley LLC, July, 2011.

[7]   Mitchell, T. D., Davis, M. W, and Kerr, S. R. III., U.S. Patent 5,164,461, to General Electric Company (November 1992).

[8]   Gray, T. E., Kunselman, M. E., Palmer, R. A., and Shulz, W. J. Jr., U.S. Patent 5,248,715, to Dow Corning Corporation (September 1993).

[9]   Gray, T. E., Kunselman, M. E., and Palmer, M. A., U.S. Patent 5,364,921, to Dow Corning Corporation (November 1994).

[10] VanWert, B., and Wilson, S. W., U.S. Patent 5,270,425, to Dow Corning Corporation (December 1993).

5 Testing adhesive bonds

There is always a question of how to determine if you have a good bond. How do you test a bond? How do you determine if it is good? The classic answer is: you break the part and if it breaks somewhere other than at the bond, it is a good bond. ASTM D429 covers "Standard Test Methods for Rubber Property – Adhesion to Rigid Substrates." This specification concerns itself with testing elastomer bonds. There are a number of test methods with various geometries listed in the specification. Adhesive failure is grouped into classifications with shorthand abbreviations to denote each type:

R indicates cohesive failure in the rubber substrate.

RC indicates the failure occurred between the rubber and the cement (topcoat).

CP indicates that the failure occurred in the interface between the cement and the primer.

CM is used to indicate failure occurred between the adhesive/primer and the metal.

Occasionally, one sees cohesive failure within the primer or within the adhesive that is so noted. **Coh** is often used as the abbreviation for this seldom-seen condition.

It is common for failure to occur in a variety of modes and the rough percentage of each (approximated visually) is recorded, for example, R 25%, RC 75%.

Four methods from ASTM D429 are commonly seen in the literature: Methods A, B, C, and F (Figure 5.1).

For Method A, a thin sheet of rubber 3.2 mm thick is bonded between circular metal plates with a diameter of approximately 40 mm. The rubber section is constrained in two dimensions, which puts it into triaxial tension. The thin rubber layer usually causes the rubber to fail in triaxial tension (Figure 5.2) as evidenced by the multiple pockmarks where failure initiates. Method A has fallen from popularity because it seldom fails at the adhesive layer and comparison with other tests reveals it to be much less able to discriminate between bonding adhesives. It is deemed by many to be more a test of the rubber strength than the adhesive strength so it gives only limited data about the quality of the bond. Results are expressed in megapascals (MPa) based on the cross-sectional area of the specimen.

https://doi.org/10.1515/9783110658996-005

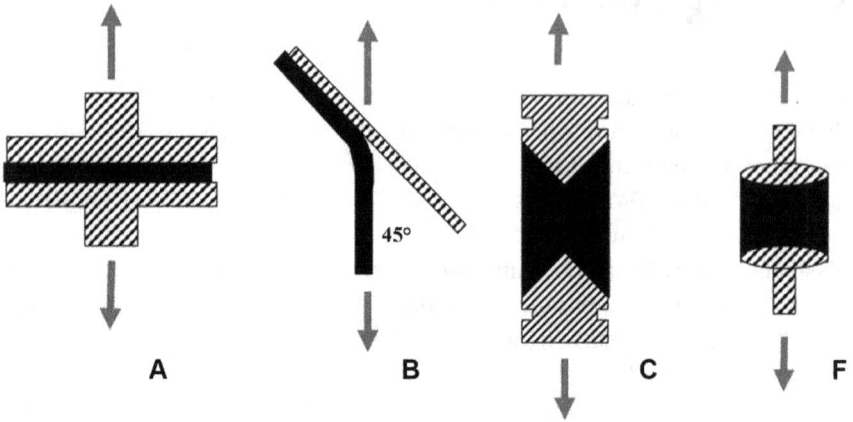

Figure 5.1: Cross sections of ASTM D429 specimens Methods A, B, C, and F.

Figure 5.2: Triaxial tension failure in the rubber in Method A specimen.

The Method B test involves peeling a one-inch wide rubber strip from a metal or plastic coupon at either a 45° or 90° angle (Figure 5.3). This creates a tear propagation rather than a tensile break. The 90° angle is standard in the ASTM procedure but the 45° angle is also allowed. The ASTM D429 procedure states, "Experience indicates a lower force is obtained for the 45° than for the 90° angle, and also a break is significantly closer to the bond surface." The 45° angle is considered by many experts to be the more stringent test because there are fewer tensile breaks of the rubber observed with this method.

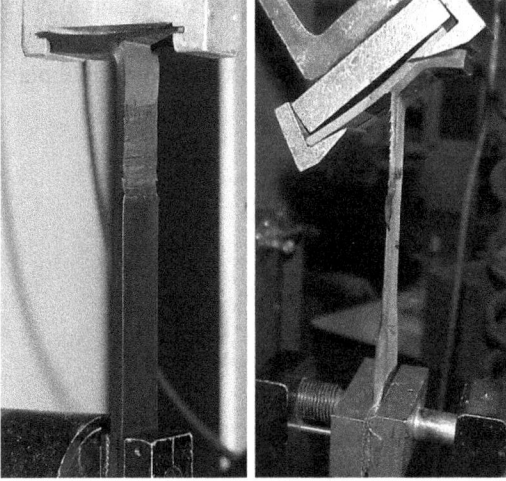

Figure 5.3: Ninety degree and forty-five degree angle tests.

Figure 5.4 shows the specimen as it is mounted in the test fixture and as it is placed under load. The tear propagates up the specimen and the peak tearing load can be recorded in Newton per millimeter (N/mm) or in pounds per linear inch (PPI).

Figure 5.4: Specimen mounted for testing and as it goes into tension.

Method B is a popular test for several reasons. It requires only a simple compression mold (Figure 5.5) and readily available compression presses, and the coupons are inexpensive. They are easily obtained in many different rigid plastics or metal alloys.

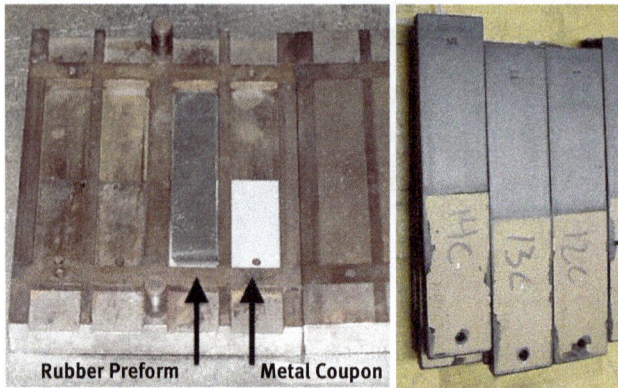

Figure 5.5: Method B mold and specimens.

ASTM Method C specimens use two conical metals bonded with the metal tips facing each other (Figure 5.6). When pulled, the sharp tips create a high stress area and failure initiates at the sharpened point. The diameter is 1.0 inch or 25 mm, and the separation of the tips is 11.5 mm.

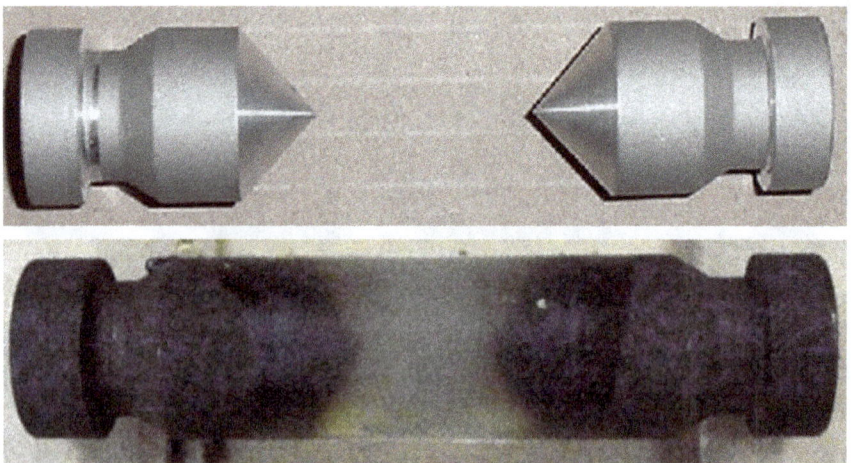

Figure 5.6: Conical metals and a bonded specimen with a clear elastomer.

After failure is initiated, it proceeds as a tear front progressing up the conical metal specimen from the tip where it originates. Because the surface area of the tearing path is continually increasing, the loads continue to increase until the specimen is completely broken. The peak load is reported along with the failure type. Method C conical specimens tend to have lower standard deviations than Method B coupons. However, the metals are expensive because they have to be machined and while they can be salvaged, the face occasionally needs to be remachined to keep the points sharp.

A recently introduced test specimen is the Method F or buffer specimen (Figure 5.7). It was designed from an actual vibration mount, so it more directly resembles a commercially bonded part than the other ASTM specimens.

ASTM D429 F Drawing

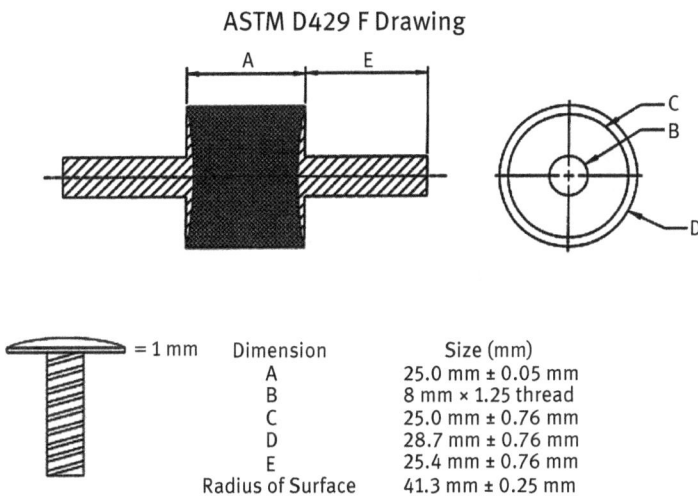

Dimension	Size (mm)
A	25.0 mm ± 0.05 mm
B	8 mm × 1.25 thread
C	25.0 mm ± 0.76 mm
D	28.7 mm ± 0.76 mm
E	25.4 mm ± 0.76 mm
Radius of Surface	41.3 mm ± 0.25 mm

= 1 mm

Figure 5.7: Drawing for the Method F specimen.

The metal specimens are cast and so they are less expensive to produce than Method C conical specimens. The molded specimens are screwed into a fixture (Figure 5.8) and then they are pulled in tension (Figure 5.9). The tensile break load is recorded along with the failure type. This test is favored by many because it closely resembles the actual part geometries. It also has lower standard deviation of results.

Two other ASTM methods should be noted. Method D is specifically designed for postvulcanization bonding. It is used to determine the strength of a bond formed between vulcanized rubber and a substrate. Method D uses a circular disk of vulcanized rubber assembled between two parallel metal plates.

Figure 5.8: Method F specimen screwed into a test fixture.

Mounted for Test   **Nearing Failure**

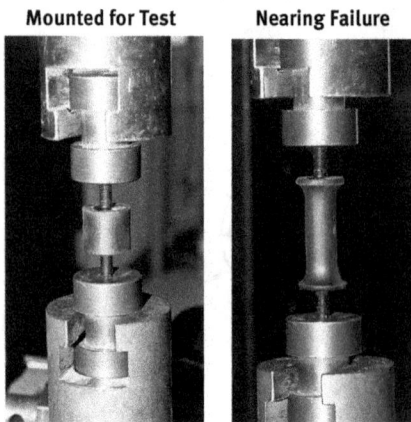

Figure 5.9: Method F specimen being pulled in tension.

The metal plates are coated with an adhesive and the assembly is placed under pressure and exposed to heat for a time and temperature sufficient to react with the adhesive. The rupture force and mode of failure are recorded after being pulled in an axial direction.

ASTM Method E is a 90° stripping test designed to simulate the adhesive strength of rubber tank linings. Somewhat similar to the method B test, the rubber strip is separated at a 90° angle. Unlike Method B where the rubber is compression molded under pressure, the rubber for Method E is cured by steam pressure or otherwise subjected to the conditions anticipated for use for bonding the tank lining in actual application.

There are many non-ASTM tests used by industry. Primary adhesion is only a first indicator of bond integrity. Having a good primary bond does not guarantee that the bond will withstand degrading environmental conditions.

One test that has been used traditionally to evaluate secondary adhesion (environmental robustness) is a stressed test. This can be run in heat, boiling water, or salt-spray (Figure 5.10).

Figure 5.10: Stressed Method B coupon test ready for heat or salt spray.

In the stressed boiling water test, ASTM D429 Method B coupons are stressed with a 2.2 kg weight while submersed in boiling water. The weight is attached to the "tail" of the bonded part via a pulley. The angle formed between the stressed "tail" and the metal coupon is 135° (Figure 5.11). The tank is then filled with boiling water and the specimens are monitored until failure.

The Method F buffer test also has an advantage that the specimen can be put into tension (Figure 5.12) and exposed to aggressive environments like heat or salt spray (Figure 5.13). The specimens are observed at intervals, and the time to failure is recorded (Figure 5.14).

Despite testing adhesives using standard ASTM methods and standard test specimens, it is a common manufacturing practice to break one part in tension or shear from a production lot part to verify the bond quality of that manufacturing lot (Figure 5.15).

For decades, there have been those who advocated that the only criterion that mattered is the bond appearance with 100% rubber being the only acceptable response. Discussions and occasional conflicts occur in connection with the writing, enforcing, and interpreting specifications for such bonds.

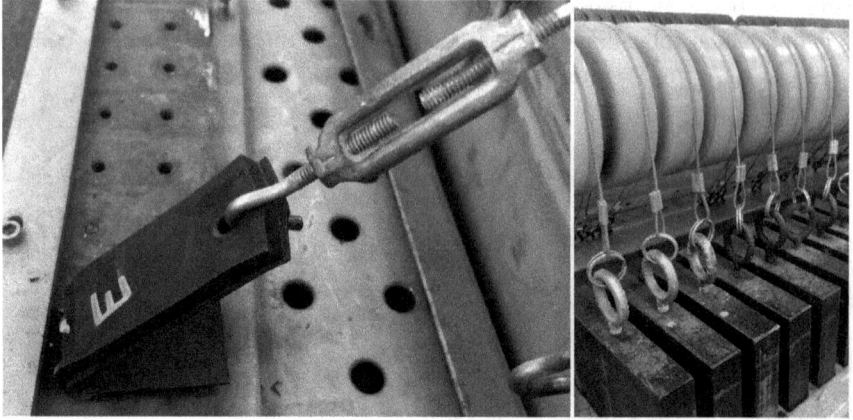

Figure 5.11: Stressed Method B coupon test ready for boiling water (left) and the weights (right).

Figure 5.12: Method F buffer specimens mounted in a fixture and placed in tension.

From an engineering point of view, the best known and perhaps simplest criterion for good bonding is readily demonstrated in welding practice. As long as any forced break of a welded assembly always occurs away from the weld and in one of the parent substrates, the weld can be considered fully accept-able. Clearly, the resistance of any assembly to applied stress cannot exceed that of the principal materials in the load path, so failure in the parent metal is the best that can be expected. This demonstrates that the weld itself is stronger than the main mass of metal, although the true strength of the weld itself is not readily determined, neither it is of particular importance to the engineer.

This same criterion has often been applied to rubber-to-metal bonding starting in the early part of this century. This meant that for many years it was reasonably valid to judge rubber bonds by their appearance only. A totally rub-ber-covered metal piece after bond rupture confirmed the original bond quality, and a clean metal surface strongly implied a serious problem with the bonding

Figure 5.13: Stressed buffer specimens placed in a salt spray chamber.

Figure 5.14: Stressed buffer specimens showing bond failure in salt spray.

process. The practice of writing specifications calling for destructive bond testing with a minimum 95% rubber coverage of the metal surface became common. In many cases, this led to major difficulties for rubber molders, who were held to visual standards for failed bonds even though the rubber was clearly improved and the bonds were stronger. Many lengthy discussions over what constituted a torn rubber surface layer took place. For example, photomicrographs did show that a seemingly clean metal surface could actually have a very thin film of silicone rubber covering it, and some specifications were revised to allow for a minimal thickness of torn rubber layer.

Figure 5.15: A bonded multilayer part (top) being broken in tension (middle) to evaluate bond quality and showing 100% rubber retention after break (bottom).

Prior to 1997, no detailed experiments had been described in the literature comparing Methods A, B, C, and F. At that time, Halladay and Del Vecchio undertook to run such a comparison [1]. There was also contention at that time as to whether the newly introduced aqueous adhesive systems were as good as the solvent-based systems used previously. For the study, two polymer types were chosen: natural rubber which is strain crystallizing and styrene–butadiene rubber which is not. Two formulas of very different levels of polymer content were developed of each polymer type. A high rubber hydrocarbon content compound was used to represent a high-quality compound, and a lower hydrocarbon content compound was used to be indicative of a much cheaper compound extended with filler and oil. All four compounds being approximately 50 durometer. Two solvent-based and two aqueous adhesive systems were chosen, all four systems employing both a primer and a cover coat. The two solvent-based adhesives and the two aqueous adhesives were selected such that each of them is based on different chemistries.

Once the testing to break was completed, all specimens were examined and a rating method for percent rubber retention and mode of failure was developed so that the results could be analyzed using statistical design methodology. Rubber retention was rated in six levels (Figure 5.16) as follows:

1. Less than 5% rubber remaining on surface
2. More than 5% but less than a third of the surface retains rubber
3. Between one-third and two-thirds of the surface is covered
4. Over two-thirds but less than 95% of the surface is covered
5. All the surface is covered with a thin layer of rubber
6. All the surface is covered with a thick layer of rubber

Figure 5.16: Rubber retention rating scale from 1 (upper right) to 6 (lower left).

The experiment was run in a full-factorial design of 16 runs (Figure 5.17). The formulations are shown in Figure 5.18.

Compound	Polymer	Quality	Adhesive	Type
A	SBR	Low	205/220	Solvent
B	NR	Low	205/220	Solvent
C	SBR	High	205/220	Solvent
D	NR	High	205/220	Solvent
A	SBR	Low	8001/8200	Water
B	NR	Low	8001/8200	Water
C	SBR	High	8001/8200	Water
D	NR	High	8001/8200	Water
A	SBR	Low	205/253X	Solvent
B	NR	Low	205/253X	Solvent
C	SBR	High	205/253X	Solvent
D	NR	High	205/253X	Solvent
A	SBR	Low	8001/8560	Water
B	NR	Low	8001/8560	Water
C	SBR	High	8001/8560	Water
D	NR	High	8001/8560	Water

Figure 5.17: Full-factorial design.

Compound Indentifier	A	B	C	D
Compound Type	SBR	NR	SBR	NR
	Low RHC	Low RHC	High RHC	High RHC
NR CV60		100.00		100.00
SBR	100.00		100.00	
Zinc Oxide/Stearic Acid	7.00	7.00	7.00	7.00
Antidegradants	2.00	2.00	2.00	2.00
Aromatic oil	22.00	45.00	10.00	
Carbon black	5.00	30.00	25.00	25.00
Pptd. Calcium Carbonate	50.00	75.00		
5 Micron Silicon Dioxide	50.00	200.00		
Sulfur	2.00	1.50	1.50	1.50
CBTS		0.70		0.70
MBTS	1.00		0.80	
TMTM	0.50		0.50	
% Rubber Hydrocarbon	41.8%	21.7%	68.1%	73.4%

Figure 5.18: Formulations.

One comparison was to look for differences between the data from solvent-based bonding agents versus water-based, and none of the responses showed a statistically significant difference indicating that aqueous systems are as good as the solvent-based systems. The test specimen geometry impacts the ability to discriminate between the bond strength of different formulations and adhesive systems. Of the methods used, Method A is the least discriminating and the other tests should probably be used preferentially, depending on what kind of application is being evaluated. The Method F specimen had the most consistent ability to discriminate differences in responses. Individual adhesive combinations will at times interact with the compounds, in that sometimes two different adhesives will have very similar results with one compound but contrasting results with another.

Performing a simple correlation study between Method F buffer bond strength and retained rubber shows that for these four compounds and four adhesive systems, there is actually a general negative correlation (Figure 5.19). While this does not prove that such a positive relationship never exists, it does demonstrate that the simple assumption of a positive correlation is unjustified.

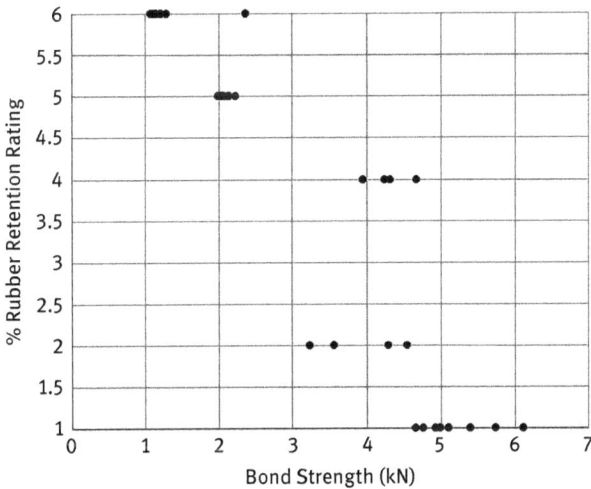

Figure 5.19: Rubber retention rating versus bond strength.

Another question that is reasonable to ask is whether there is a correlation between the results of the Method F buffers and the Method B coupons. A plot of the test results shows that there is poor correlation (Figure 5.20).

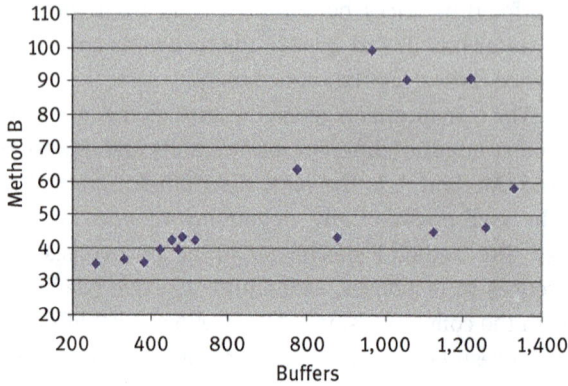

Figure 5.20: Method F buffers versus Method B peel shows poor correlation.

Most importantly, this seminal study gave proof for the first time that the appearance of the failed bond surface does not have a positive correlation with bond strength, and the use of visual standards alone in preference to force-to-failure measurements is inappropriate in evaluating bond quality (Figure 5.21).

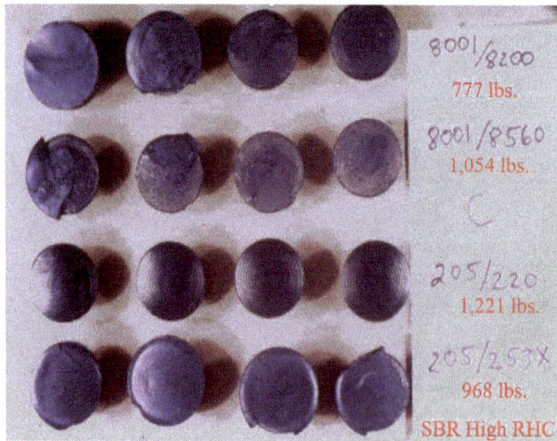

Figure 5.21: The highest bond strength for this SBR compound bonded with four different adhesive systems was achieved with the Chemlok 205/220 adhesive system that had 1,221 lbs. break force but 0% rubber retention at break.

If the force measurements are lower than expected, bond appearance can be an important clue as to why the break values were low and they can tell us which part of the bonding operation we should focus on for the cause and corrective action.

Reference

[1]   Rubber-Metal Bonding Studies Using Designed Experiments. Del Vecchio, R. J. and
       Halladay, J. R., 152nd Meeting of the Rubber Division ACS. October, 1997. (also published
       in Rubber and Plastics News, 3/9/98).

6 Substrate preparation

Most substrates, as received, are unsuitable for bonding without surface preparation. Unprepared metal surfaces have a natural metal oxide layer. Metal oxides result when oxygen combines with a metal and is evidenced by red rust on iron and steel alloys, or the white powdery corrosion sometimes seen on aluminum. It may be nearly invisible on aluminum, titanium, and stainless steel. Metal oxides are usually of low strength, and long-term adhesion is generally poor when paint or adhesive is applied directly to the metal's naturally occurring oxide layer.

Other factors can cause an uncontrolled surface layer that is unsatisfactory for bonding. One type includes any of the substances that react with the metal. In addition to oxygen, other common reactants include sulfur and chlorine. A second type of materials capable of causing uncontrolled surface layers are substances that become physically attached, such as oils, evaporated fuels, silicones, and solid particulates. This includes inorganic contaminants including weld smut or scale, shop dust, metallic residue, and solder or flux.

Organic contaminants often form oily films, and they can come from the drawing lubricants used in metal manufacturing process, corrosion inhibitor oils or residue, coolant or cutting fluid used in machining, fingerprints from improper handling, or silicone from hand lotions and mold release sprays. Paraffinic-based oils can be particularly difficult to remove.

The objective of surface preparation is to remove the naturally occurring, impure metal oxide surface layer to expose a higher energy surface of fresh metal atoms and to obtain a controlled surface chemistry and surface structure to which adhesion will persist under long-term, harsh environmental conditions. It also serves to provide consistent and reproducible adherend surface. A strong initial bond (referred to as primary bond) is not often indicative of how well the bond will last when exposed to service environments. A part can show an excellent primary bond immediately after molding and still fail due to other factors such as environmental conditions or fatigue cycling. Examples of mild to harsh environmental conditions are shown in Figure 6.1.

The selection of the pre-bond surface treatment is dependent on the application, the environmental service conditions, and manufacturing cost sensitivity. A number of industrial cleaning processes have been used over the years. These include solvent degreasing, solvent washing or wiping, alkaline cleaning, acid etching, and grit blasting. Solvent degreasing is quite effective at removing organic soils, but it is highly regulated. Acids remove rust and

https://doi.org/10.1515/9783110658996-006

Coastal Region
Dry Indoors

Outdoors
Dry Climate

Coastal Region
Damp Climate

Coastal Region
Damp Climate
Polluted Air

Down-hole
Drilling

Outdoors
Damp Climate

Coastal Region
Dry Climate

Marine Applications
Non-submerged
Salt Water

Near
Volcanic
Activity

Indoors

Marine Applications
Non-submerged
Fresh Water

Marine Applications
Submerged
Fresh Water

Marine Applications
Submerged
Salt Water

Mild → Extremely Harsh

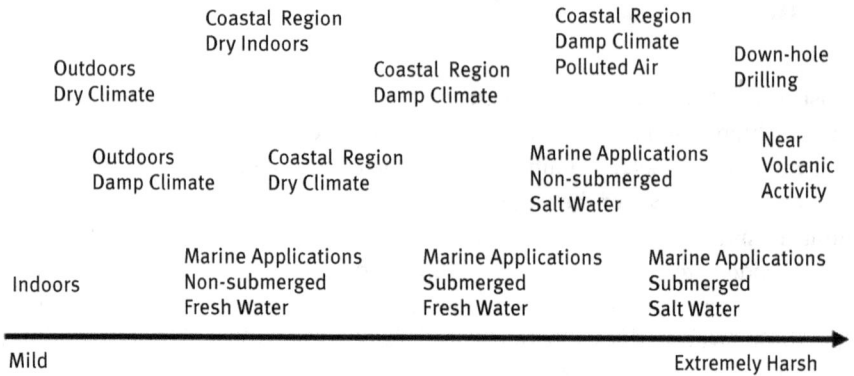

Figure 6.1: Environmental conditions ranging from mild to harsh.

scale and some organic soils. Alkaline cleaning removes organic soils and minor rust and scale.

Solvent degreasing used to be the standard for cleaning metal, but it is seldom used today because of environmental regulations. The solvents that were used were either 1,1,1-trichloroethane or 1,1,2,2-tetrachloroethylene also known as perchloroethylene or just perc. Solvent degreasing can be divided into two types, immersion and vapor. Immersion is, as it sounds, complete immersion of the part in a hot solvent bath where the oily residues are extracted by the liquid solvent. Vapor degreasing involves suspending the part in the vapors of boiling solvent. The vapors condense on the cold metal surface and run off, carrying away particulates and residues. This removes virtually any coolant, lubricant or protectant residue rapidly. It leaves a surface residue with a chemical affinity for primers which enhances the bond. The surface residue is resistant to atmospheric corrosion which gives increased layover times. Vapor degreasing may more effectively remove substances that have been forced deep within the natural oxide layer as a result of metal forming operations.

Solvent wiping involves flushing or wiping the metal surface with a liquid solvent. It is much less controllable since only the operator passes judgment as to when the surface is chemically clean. The surface can become contaminated with spent solvent, dirty wiping rags, or soiled handling equipment. It is labor intensive and generally used only when absolutely necessary.

Alkaline cleaning is usually based on either sodium or potassium hydroxide. Either of these strong, water-soluble bases reacts with water-insoluble ester oils, converting them into water-soluble organic salts and alcohols. This is known as saponification, and it is the same reaction used in making soap when animal fats and lye are mixed. The net result is that some oils are chemically changed, forced into solution, and are effectively removed from the surface of the part.

The alkaline cleaning solutions contain surfactants and detergents. The detergent has both wetting and emulsifying properties. Detergents are typically long organic molecules with a hydrophobic end and a hydrophilic end. The hydrophobic end dissolves the oils and the hydrophilic end (negatively charged) dissolves in water. This causes the oils to, in effect, dissolve in water. These components of an alkaline cleaning solution remove particulate matter and oils. Surfactants and detergents change the electrical properties of the interaction between the surface and particulate contaminants in a manner that reduces surface tension and electrostatic attraction, causing unwanted particles to dissociate from the surface. The water-soluble detergent or surfactant interacts with hydrocarbon oils forcing an association between the two on the molecular level. The intermixture of oil and water molecules form tiny domains called micelles which disperse readily in water. The dissolution of oil into a water solution is enabled, and hydrocarbon oils are removed from the surface of the part (Figure 6.2).

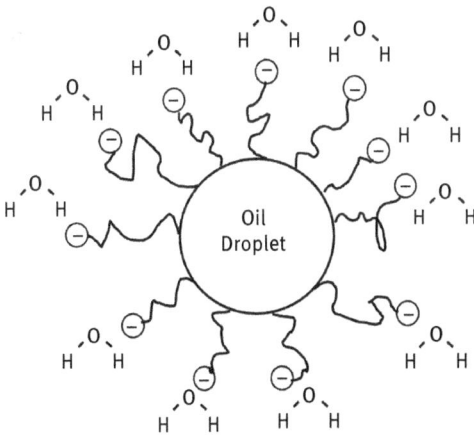

Figure 6.2: A soap micelle in water.

Alkaline cleaning usually performs adequately, and it is environmentally less offensive than most alternatives. It is cost effective and reasonably controllable. However, there are some drawbacks to alkaline cleaning as compared to solvent degreasing. Unlike solvent degreasing, alkaline cleaning does not provide for a reduction in the rate of atmospheric corrosion. It does sometimes cause steels to rust when immersed in the water solution. It does not leave a residue with an affinity for primers. Aluminum and zinc alloys can be rapidly consumed by high-pH caustic solutions. Even if appropriate pH adjustments have been made for applications involving aluminum, certain aluminum alloys can still be consumed, and it can cause corrosion problems with all 2000 & 7000 series aluminum alloys.

Grit blasting is commonly used to remove surface oxidation layers and to prep metallic parts for chemical treatments. Some of the abrasive particles used include steel grit and shot, aluminum oxide, glass beads, plastic beads, silica sand, mineral slag, walnut shell, and corn cob. Blasting removes surface coatings and contaminants. It increases surface roughness which increases the surface area. Blasting can be controlled by the type, size, and hardness of the abrasive and the pressure, distance, and angle of the blast nozzle. Blast profiles are measured in either mils or microns. A mil is 1/1,000th of an inch and a micron is 1/25 of a mil (25 microns = 1 mil). Precise profiles are not possible due to varying sizes in the abrasive. Aluminum oxide grit between 25 and 80 grit particle size is generally recommended. When coating over a blasted profile, it is important that the coating is applied so as to cover the peaks of the blasted profile (Figure 6.3). When surface profiles exceed coating thickness, the peaks may protrude nearly through the adhesive, causing high-stress areas in the very thin coating at the top of the peaks that can lead to bond failure.

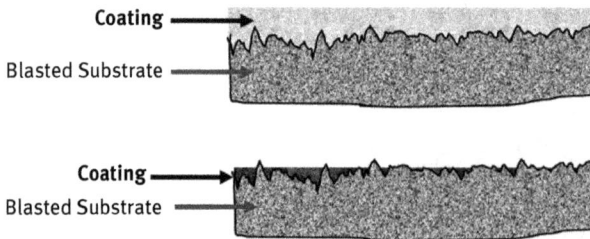

Figure 6.3: The coating should cover all of the peaks of the blasted profile, as shown in the top illustration.

If in doubt of where to start, a one to three mil blast profile using a 40 grit aluminum oxide blast is a recommended starting point, being careful not to overblast. For a properly blasted surface, one can still observe individual craters. In an overly blasted surface, the metal has become smeared, and individual craters can no longer be distinguished (Figure 6.4).

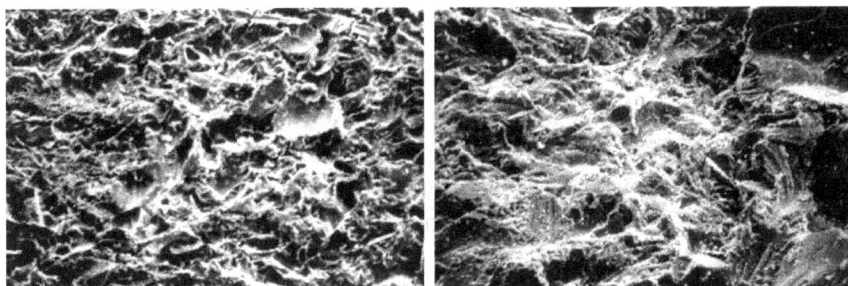

Figure 6.4: Properly blasted (left) and over-blasted (right) at 400-X magnification.

In most cases, an alkaline clean, followed by a blast, and a second alkaline clean is a robust procedure. The first alkaline clean removes oils so that they do not contaminate the blast media. The blast removes the oxide layer, and the final alkaline clean removes dust from the blast operation and any oils that may have been introduced from the blast media. It is recommended to apply the adhesive between one to two hours after cleaning.

Another treatment, acid etching, can be used to remove inorganic soils. Some of the acids used include phosphoric, hydrochloric, sulfuric, and certain organic acids like acetic, citric, gluconic, oxalic, and formic.

Carbon steel is often phosphatized to improve environmental resistance. Phosphating involves immersing the metal in a phosphoric acid solution to form a soluble phosphate salt. The soluble salt is then converted into an insoluble metal phosphate crystal which is bonded to the metal surface. The actual process consists of the following steps:
- Hot caustic clean
- Water rinse
- Phosphoric acid pickle
- Water rinse
- Zinc phosphate bath
- Water rinse
- Sealer (optional)
- Hot air dry

The rinsing removes chemicals and loose soils. It provides separation between alkaline and acid processes so that cross-contamination is kept to a minimum. This requires an overflow in the tank and regular dumping and cleaning of the rinse tanks. Parts should always remain wet between the process baths.

The insoluble metal phosphate provides a physical barrier against moisture and mild corrosion. It also provides a base for paint, adhesives, or oil and rust preventative treatments. The phosphatized surface enhances lubricity and resistance to wear. The two most commonly used phosphate coatings are iron and zinc.

Iron phosphate has a lower coating weight than zinc and is recommended to be applied between 10and 50 mg/square foot. It is usually applied using either iron molybdate or chlorate and is organically accelerated. It performs well with rubber adhesives, is excellent for swaging, and is more environmentally friendly than zinc.

Zinc is divided into heavy zinc, medium zinc, and calcium-modified zinc phosphatize. Medium zinc runs a coating weight between 400 and 1,400 mg/square foot with crystal size between 18 and 40 microns. The coating weight for heavy zinc runs a coating weight above 1,400 mg/square foot and crystal size greater than 40 microns. Heavy zinc is unsatisfactory for elastomer bonding. The large plate-like crystals are brittle. They fracture and break away from the metal surface when stressed. Most preferred for bonding is the calcium modified zinc with a coating weight between 150 and 400 mg/square foot with crystal size between one and seven microns (Figure 6.5). It is important to note that zinc phosphatize should not be used for service temperatures greater than 275 F (135 C) as it dehydrates the crystal and turns it to powder. The use of a dichromate sealer enhances the under-bond corrosion resistance.

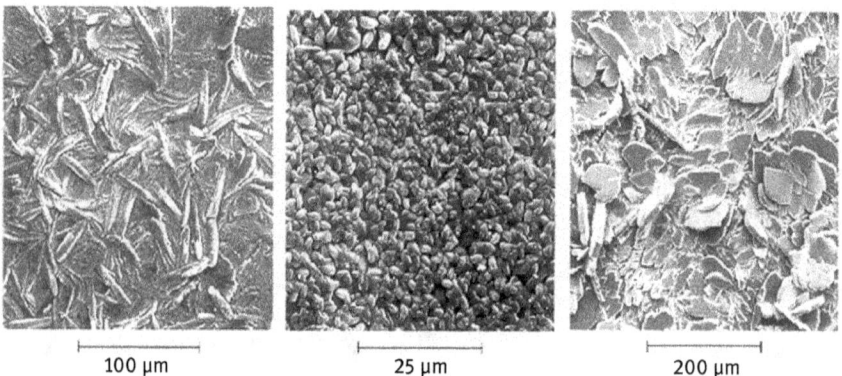

| 100 μm | 25 μm | 200 μm |

Figure 6.5: Medium zinc-calcium-modified zinc heavy zinc.

Smutting is a potential problem in any pickling process or in a chemical bath that has the potential for consuming metal substrate. Smut is a loosely adherent, usually gray to black, fine surface residue. The primary causes of smutting include excessive immersion time, over-tolerance of the chemical concentration in the bath, or to high of a temperature in the bath. Smutting can be removed by hand wiping or with a chemical treatment.

Although stainless steels are corrosion-resistant, they are not completely impervious to rusting due to ferrous smudge on the surface. Stainless steel should be blasted with aluminum oxide or glass beads and not with steel grit to prevent galvanic reaction due to contamination with dissimilar metals. Passivation, an acid etching process using nitric or citric acid-based baths, is used to deactivate the surface and make it more resistant to corrosion (Figure 6.6). Nitric acid passivation removes trace quantities of free iron (ferrous smudge) and converts exposed chrome and nickel on the surface to their oxides, restoring a passive oxide layer. It is the recommended surface preparation for stainless steel. Passivation is not effective in treating forgings and castings because of the nature of their surfaces. Another etchant that can be used for preparing stainless steel for bonding is Arcal Stainless Steel Etch #12, manufactured by Arcal Chemicals, Inc.

Titanium is often etched with a hydrofluoric acid etchant (Figure 6.6). Primer or adhesive should be applied within eight hours of etching.

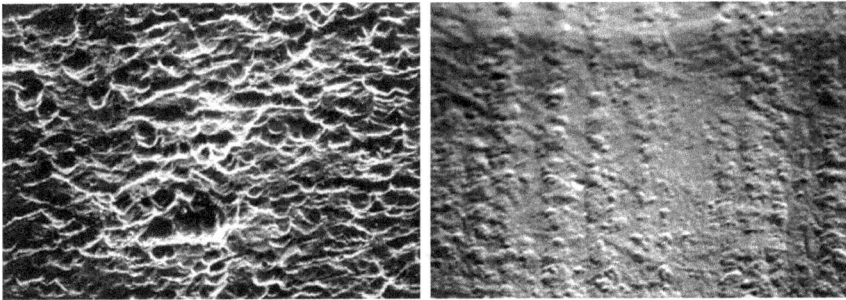

Figure 6.6: Etched surfaces, stainless steel (left) and titanium (right).

A chromate conversion coating on aluminum increases corrosion resistance and improves adhesion of paints, primers, and coatings. It changes aluminum's weak, naturally occurring oxide layer to a stronger chemically resistant aluminum oxide/chromate-complex. Chromate conversion coatings can also be used on zinc, cadmium and magnesium. Unlike a phosphatize coating, a chromate conversion coating gives resistance to corrosion even when a paint or adhesive is not applied to the surface. Alodine (defined in MIL-C-5541) is a chromate-conversion coating (non-electrolytical) that is a preferred process for bonding aluminum. The process consists of the following steps:

- Hot caustic clean
- Water rinse
- Deoxidize
- Water rinse
- Chrome conversion
- Water rinse
- Hot air dry

The chromate conversion bath contains hexavalent chromium, and its use will likely be phased out over the next several decades.

It is common to anodize aluminum in preparation for bonding. Anodize is an electrochemical process performed per MIL-A-8625. It provides a high-strength coating that is extremely resistant to corrosion. Commonly used are sulfuric acid and chromic acid anodize coatings. Sulfuric acid anodize is preferred to chromic acid anodize for bonding, but unfortunately the sulfuric acid anodized surface reduces the fatigue resistance of aluminum alloys. If the aluminum is left un-sealed, apply adhesive within one hour.

For plastics, blast with glass bead or aluminum oxide under low pressure. Plastics with 10–30% glass fill aids in adhesion. Plastic surface preparation also includes corona discharge, plasma treatment, and flame treatments. These treatments clean and activate the surface in preparation for bonding. There are etching solutions available for treatment of polytetrafluoroethylene (PTFE) and polyetheretherketone (PEEK).

There are some novel methods for activating substrates so as to be receptive to bonding, which have had some attention in the past 20 years or so. The three most common methods are:
- Plasma spray
- Corona discharge
- Flame treatment

A plasma is formed when a stream of gas passes through a high-voltage electrical discharge. Molecules break down into ions, free radicals, and high-energy particles and the stream of these entities both blasts loose materials off the surface and undergoes energetic interactions with the surface that produce various kinds of extremely active chemical species there. The gases that may be used include air, but other gases are used as well, and the combinations of particular gasses in an electrical discharge will impart their own unique activation properties. Sometimes the activated surface will have only moderate reactivity and will degrade rapidly with time, but some other combinations of plasma with certain surfaces may produce high reactivity and a lengthy lifetime, measured in hours or days.

Corona discharge is related, in that gas exposed to an electrical field between two electrodes becomes activated and forms a cloud, which then impinges on and activates surfaces.

Flame treatment or spray is exactly that, a focused flame, often from propane or natural gas. It is generated, usually from a length of piping, and the substrate material is passed underneath it. Variables are the intensity of the flame, the distance from the substrate surface, and the length of time of exposure. Flame activation does not tend to have long duration and must be bonded immediately.

These methods have been very successfully employed to produce activated surfaces on a variety of materials, especially assorted plastics, which normally do not bond well to rubber. They have the advantage of not using solvents, and once installed, they tend to be much lower in cost to run. Even challenging problems like bonding cured silicone rubber to a smooth molded plastic surface have been solved with these methods.

There are various methods of checking for surface cleanliness. The simplest is a dry rag wipe, also known as the white glove test. It is basically as it sounds. Wipe the metal surface with a clean white cloth or glove. There should be no residue observed if the surface is truly clean. A second common method is the water break test per ASTM F22-65. Water should sheet off and not bead up on the surface. Analytically, one can run surface carbon analysis to see if any hydrocarbon contaminants are present.

In summary, the importance of having a clean substrate surface prior to adhesive application cannot be over-emphasized.

7 Adhesive application and use

The bonding process involves adhesive preparation, adhesive application, and molding. The robustness of the overall process is affected by the substrate topology and chemistry, and the presence of mold releases or weak boundary layers on the substrate surface. The process is affected by the elastomer chemistry, the adhesive chemistry, and the elastomer's cohesive strength. But it is also impacted by the processing parameters chosen. Process parameters include the time, temperature, and pressure chosen for the molding conditions. It includes the exposure of the substrate and adhesive to heat before the rubber has filled the mold (referred to as prebake resistance) and it can be affected by mold sweep.

A robust adhesive must have the following properties:
- It should readily wet and spread over the substrate surface.
- It should promote interfacial mixing without destroying the integrity of the adhesive film.
- It should possess sweep resistance.
- It should have adequate prebake resistance.
- It should develop a strong bond that resists hot tearing during demolding.
- It should maintain bond integrity through all post-bond operations.
- It should accommodate reasonable variation in cure temperatures.
- It should possess excellent environmental resistance.

Needless to say, this is a lot to ask from any adhesive. The adhesive selection process can be quite complex. What works for one compound may not work for a similar compound having a different formulation and what works for one specific part geometry may not work for a much larger or smaller part. The commercially available adhesives on the market change continually and each adhesive supplier has a plethora of products commercially available. The best advice is to consult with the adhesive supplier when the adhesives already in use in the plant do not work satisfactorily in a new application.

Like rubber, adhesives exhibit a cure rate. That cure rate can be dependent on the primer, the mold temperature, and the elastomer chemistry. Optimum adhesion to the rubber is obtained when the adhesive and the rubber cure at roughly the same time. Most adhesives are heat activated at around 140 °C and molding is generally between approximately 140 °C and 195 °C. Generally, the time required for curing the rubber is also sufficient to cure the adhesive. However, large cold metal components can act as heat sinks and can keep the

https://doi.org/10.1515/9783110658996-007

adhesive colder than the rubber such that the rubber cures before the adhesive. This may require a preheat on the metals and/or a longer cure time. Prebake resistance refers to the amount of time and temperature the adhesive can be exposed to before it cures. Prebake resistance varies depending on elastomer compound and the adhesive chemistry. Thin metal components may require a lower cure temperature to prevent the adhesive from being activated before the rubber fills the mold. The best action is to determine at what point the bond strength begins to degrade at a given molding temperature.

Sweep resistance refers to how easily the incoming rubber stream of rubber removes the adhesive in a transfer or injection molding operation. Sweep resistance varies depending on adhesive system, the elastomer, and the mold temperature. Sweep is influenced by the location of sprues where rubber is transferred or injected into the cavity and the rate at which transfer occurs.

The first step in adhesive preparation and application is the proper storage, mixing, and dilution of the adhesive system. Adhesive storage should avoid heat or cold. Temperatures of 20 to 30 °C are recommended. If a water-based adhesive or primer freezes, it can destroy the integrity of the system. Thawing it out will not repair the damage that has been done. Nonsilane-based adhesives and primers contain a number of active ingredients that are not dissolved in the solvent or aqueous carrier. These ingredients are merely in suspension and they settle out over time. Thus, like paint, the suspended ingredients end up on the bottom of the drum or can. The following procedures are recommended for mixing:

½ pint can: hand stir for 5 to 10 minutes
1 gallon can: hand stir and then use a paint shaker for 20 to 30 minutes
5 gallon can: hand stir and then use a paint shaker for 45 to 60 minutes
55-gallon drum: hand crank and then use an air driven agitator for 8 hours

Next, the adhesive must be diluted to the recommended viscosity. In the lab, a Brookfield viscometer provides an accurate measurement, but in production, Zahn cups 1 through 4 usually give adequate results. It is critical to use the proper solvent and to add it slowly. Adding solvent quickly may shock the system and cause gelation. Adding the wrong solvent can cause some of the dissolved solids to precipitate out of solution.

There are 3 primary methods for adhesive application, brushing, dipping, and spraying. Brush application has the maximum transfer efficiency, greater than 75%. It is labor intensive and has generally inconsistent film thickness (Figure 7.1) but is useful for small parts and for coating only certain areas on a part.

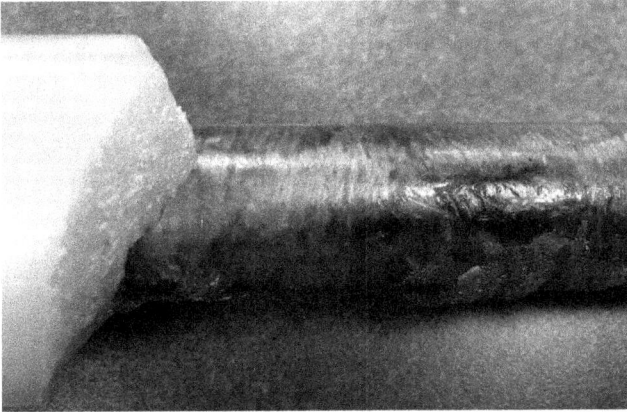

Figure 7.1: A brush-applied urethane adhesive shows inconsistent thickness which resulted in cement to metal failure.

Dip application has high transfer efficiency and consistent film thickness from part to part. All surfaces get coated whether it is desired or not. Care must be taken to avoid heavy build up at the bottom of parts and parts should be oriented so that tears and runs are directed to low stress or unbonded areas of the part. Spray gives a lower transfer efficiency (10 to 50%) than dipping or brushing but it offers the highest production throughput and gives consistent film thickness at the hands of a trained operator. Parts can be masked so that unbonded surfaces do not get coated. One must take care to use clean compressed air and to apply a uniform wet coat of adhesive. For waterborne adhesives and primers, the metal part should be heated to between 50 and 70 °C to facilitate wetting and increase the evaporative rate.

Some of the key parameters in adhesive application include avoiding exposure of the adhesive to oil, moisture, dust, and so on and using clean compressed air (water and oil free) for spraying. Avoid any contamination of primer with adhesive or of one adhesive with another. When spraying, apply a uniform wet film of adhesive and avoid dry spray or runs. Dry spray refers to the condition that occurs when too much of the solvent has evaporated before the spray reaches the part and the resulting coating arrives dry rather than wet. The primer coat should be thoroughly dry prior to applying adhesive top coat and the adhesive coat should be thoroughly dry prior to bonding. Recommended dry times are a minimum of 30 to 60 minutes at room temperature between coats or before molding.

Process control variables are film thickness and uniformity. If the coating is too thin, there will be insufficient ingredients for a good bond. If the coating is too thick, the bond may fail cohesively within adhesive layer, and the coating is more prone to adhesive sweep. Typically, one can measure the dry film thickness (DFT) using digital DFT gauges. Three detection methods are employed: beta backscatter, magnetic induction, and eddy currents. DFT gauges are readily available in either portable hand-held models or in tabletop models and they work with virtually all metallic alloys. Adhesives should be applied per the manufacturer's recommended thicknesses. Typically, primers run between 0.2 and 0.4 mils (a mil is 0.001 inches or 25 microns) and adhesive topcoats run between 0.6 and 0.8 mils. Single-coat adhesives are generally applied between 0.7 and 1.2 mils. To put this in perspective, a typical sheet of printer paper is about two mils thick!

Silane-based primers are water thin and generally applied by dipping. They are very dilute solutions in alcohol and they leave a coating too thin to be measured with a DFT gauge. In theory, the film is one molecule thick but some studies claim that in reality, they are 3 to 6 molecules thick. Suffice to say that if you can measure a film thickness, the primer is much too thick.

Handling of coated parts is important as well. Avoid touching parts while they are still wet because it will smear or even remove the coating. After parts are dry, handle them only with clean gloves. When gloves become contaminated, they should be removed and replaced. Parts should not be thrown together in a bin because it may chip off the adhesive. Parts should be stored at room temperature in a low humidity environment. Cover parts adequately for storage to avoid any airborne contaminants. Avoid storing metal parts too close to molding press since airborne mold spray can settle out on the parts. Use cemented parts within the specified shelf life. Most adhesives can be stored for up to thirty days before molding but metals coated with silane-based primers should be molded within 3 days.

Post-bond operations can have an adverse effect on bond integrity. The ability of an adhesive system in maintaining the integrity of the bond during a swaging process varies depending on the brittleness of the primer and adhesive chosen. Rough metal surfaces allow for a higher percentage of swaging than smooth metal surface. Cryogenic deflashing can damage the bond if the part is exposed to too low of a temperature for too long of a time. Corrosion inhibitors applied after bonding should be checked to ensure that they do not damage the bond over time. Post-bond plating operations can cause debonding to occur at the edges if the part is exposed to the baths for too long or if the part sees too high of a current density in the plating bath.

Postvulcanization bonding, generally referred to as PV bonding, can be accomplished with elastomeric adhesives which cure at high temperatures or with a urethane or epoxy adhesive which cures at room temperature or a moderately

elevated temperature. As with normal vulcanization bonding, the substrate preparation is required for any of the methods of PV bonding. Next, the rubber must
be prepared for bonding. At the very least, a solvent wipe with MEK (methyl
ethyl ketone) or MIBK (methyl isobutyl ketone) should be used to remove mold
release agents, dust, and oils from handling. It is preferable to blast the surface
to remove the glossy skin and expose a fresh rubber surface and then solvent
wipe to remove dust. Some people advocate slightly undercuring the rubber but
that will depend on the polymer and the cure system used. Polar rubber like nitrile and neoprene are easier to PV bond than nonpolar rubber like NR, BR, IIR,
SBR, and EPDM. Nonpolar elastomers can be treated with a solution of trichloroisocyanuric acid in a solvent (e.g., Chemlok 7701 from LORD Corporation), which
will chlorinate the surface and render it slightly polar. This is applied by dipping,
wiping, or brushing and it is required for room temperature PV bonding with
epoxy or urethane adhesives.

Only a few vulcanization-bonding adhesives will work for PV bonding.
Consult your adhesive supplier to determine which adhesives are acceptable.
For cured-rubber to metal, a primer is still required for the metal surface. For
rubber-to rubber, only the adhesive is required and it should be applied to both
rubber surfaces. Adhesives are generally applied at higher dry film thicknesses
than when doing vulcanization bonding. A good starting point is 1.0 to 1.5 mils
dry film thickness.

A fixture is required to hold the rubber in position and to apply a compressive
force. A compressive force of 10 to 40% is typically applied to the mating surfaces.
The fixtured assembly is then exposed to heat using a convection oven, an autoclave, or an induction heater. In a convection oven, 30 to 60 minutes at 150 °C to
165 °C is common. Autoclave cures of 15 to 30 minutes at 140 °C to 160 °C are
recommended. For parts with large metal components, these times may have to
be greatly extended. Let parts cool to room temperature before removing the fixturing. Induction heating can raise the metal temperature to 175 °C to 215 °C
quickly and cures may be complete in about one minute.

Consult the adhesive supplier for handling recommendations for epoxy and
urethane adhesive. These adhesives do not require the compression that the
high temperature adhesives require. In fact, too much pressure will squeeze the
epoxy adhesives out and the bond line will be too thin. A bond line of 10 to
30 mils is recommended for epoxy and this can be controlled with the addition
of around 1% glass beads. Fixturing of the assemblies is required until handling
strength is achieved, typically 24 hours at room temperature or 4 to 5 minutes
at 150 °C. Urethane adhesives are similar in use to epoxy adhesives except they
are not recommended for bonding to bare metal.

7.1 Troubleshooting in the factory

Troubleshooting bond problems is an important part of normal production routine. As stated previously, when high break values are obtained, bond appearance does not correlate well with bond integrity. Failure can occur in any phase or in mixture of phases. However, when the bond break values are low, the mode of failure can be enlightening as to where the problem originates. ASTM D729 defines the types of visual failure modes (Figure 7.2).

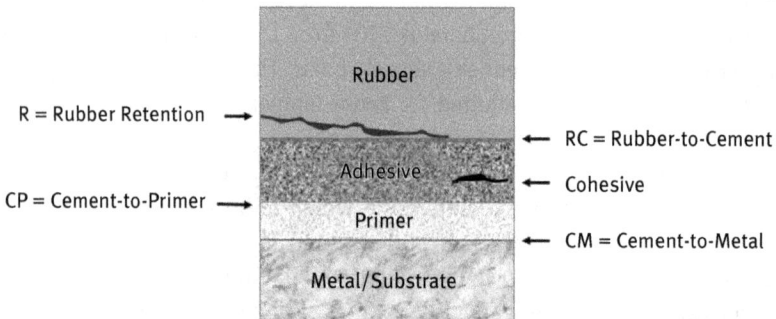

Figure 7.2: Failure types per ASTM D729.
R indicates cohesive failure in the rubber substrate.
RC indicates the failure occurred between the rubber and the cement (topcoat).
CP indicates that the failure occurred in the interface between the cement and the primer.
CM indicates failure occurred between the adhesive/primer and the metal.
Coh is not a recognized ASTM D729 failure mode, but it indicates cohesive failure within the primer or within the adhesive layer.

The first step in troubleshooting bond problems is to determine the failure mode and to review any material or process changes that have taken place.

Rubber to cement (RC) failure (Figure 7.3) is quite a common occurrence and there are numerous causes of rubber to cement failure. It probably accounts for more than 50% of the bond problems encountered in production.

One common cause is a poorly mixed adhesive topcoat, either due to a lack of initial agitation or to lack of agitation over time during use which allows settling to occur. Lack of agitation causes many of the active ingredients to settle to the bottom of the can or drum instead of being dispersed in the adhesive being sprayed. RC failure is also likely when the adhesive coat is sprayed too thin so there is an inadequate amount of the reactive agents in the bond layer. In Figure 7.3, we can observe that the adhesive is extremely thin because we can see that the metal is not well covered. RC failure can occur when the cover

Figure 7.3: RC failure is most likely caused by too thin of an applied adhesive layer.

coat is incompatible with the rubber being bonded. If the rubber cures significantly faster than the adhesive, one can expect to see RC failure. The rubber may cure too quickly because the rubber section is very thin and cures quickly, or because the rubber was formulated with ultra-accelerators that make it cure very fast, or possibly because the rubber is over-aged and has already begun to cure (scorch) while sitting on the shelf. The adhesive may be under-cured because of heavy metal inserts that are cold and act as a heat sink or because there is poor insert-to-mold contact which limits the inserts heating up in the mold. Figure 7.4 is a heavy steel part with only a thin layer of rubber bonded to it. There is both CM and RC failure visible but only in the center of the part. The outside edges exhibit good rubber-tearing bonds which implies that the center of the metal insert never heated up sufficiently.

Figure 7.4: The center of the metal insert shows a combination of RC and CM failure, indicating that it was too cold in the center to adequately react with the adhesive.

RC failure is also likely if the adhesive topcoat has started to cure or is fully cured before the part is placed in the mold. This can occur when the metal part has been preheated for too long, or when the insert loading time takes too long at high mold temperatures, or with small inserts that heat up too quickly and cure before the rubber completely fills the mold. In Figure 7.5, the glossy appearance on the small center insert suggests that the adhesive was fully cured before the rubber filled the cavity.

Figure 7.5: The shiny surface of the adhesive on the small metal insert suggests that the adhesive was cured before the rubber filled the mold.

RC failure can be caused when the rubber injection or transfer time is too slow. Flow lines in the rubber are often an indication of slow transfer time (Figure 7.6).

Figure 7.6: Flow lines in a bonded rubber part may indicate scorched rubber or a slow transfer.

If the mold pressure is too low, such that the rubber is not under adequate pressure when both the rubber and adhesive are curing, RC failure may result. This can occur when there is insufficient rubber to fill and pressurize the mold cavity creating a no-fill (Figure 7.7) or when there is poor cavity cut-off allowing the mold to leak. Poor cavity cut-off lets the rubber continue to leak out of the cavity and the rubber movement during cure often results in edge-bond problems (Figure 7.8).

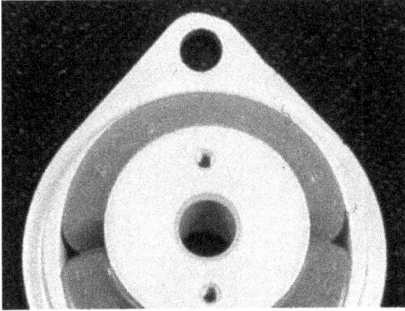

Figure 7.7: Example of a non-filled part.

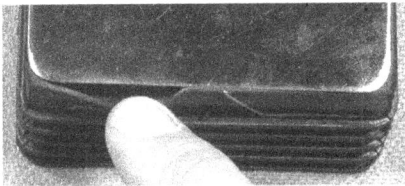

Figure 7.8: Example of edge-bond problems.

In other cases, RC failure may be due to molding at a temperature that is too low to activate the active ingredients in the adhesive (below 140 °C). If it occurs only in certain cavities, RC failure might be occurring because of nonuniform heating of certain cavities.

RC failure can also occur because of migration of certain compounding ingredients from the rubber, such as plasticizers and antiozonants. It might also occur because the coated metal parts have been contaminated with oil, finger prints, or mold releases. Using outdated or expired adhesives or using coated metal parts that have exceeded the recommended layover time can contribute to RC failure. Isocyanate and silane-based adhesives are moisture sensitive, so it is particularly important to bond those adhesives within the recommended 3-day layover time.

Cement to metal (CM) failure (Figure 7.9) is the next most common occurrence and probably accounts for about 30% of the bond problems encountered in production. In this case, the primer of the 2-coat adhesive system fails to the metal surface exposing bare metal.

CM failure is often due to poor substrate preparation, that is, the metal is not clean. This can be due to problems in either the mechanical or chemical treatment processes. On the mechanical side, it may be because of incomplete blasting or because the blast residue was not cleaned from surface. CM failure can occur if the metal surface becomes contaminated with oil before, during, or after the blast. It could also happen because too much time elapsed between the blast and the time the primer is applied, allowing an oxide layer to reform (Figure 7.10).

Figure 7.9: Cement to metal failure showing the gray primer on the rubber surface.

Figure 7.10: This CM failure on aluminum could be due to failure to remove blast residue or because the adhesive was applied over a weak oxide layer.

In the chemical processes, CM failure could be due to incomplete deoxidation of the metal or because a conversion coating was too thick. It may be attributed to improper rinsing or, as with the mechanical treatment, too long of a layover time between the chemical treatment and the time the primer and/or adhesive was applied. CM failure can occur if smutting occurred in the chemical treatment lines and it is not removed prior to primer/adhesive application. CM failure may occur when the primer is improperly mixed, improperly diluted (wrong solvent), has become contaminated, or is overaged. In the spray application process, dry spray, too thin of a primer coating, or contaminated atomization air may be the contributing factor. Cement to metal failure can result during the molding process if the primer and adhesive are swept off the metal surface. This condition can occur when the sprues are located near the metal surface (Figure 7.11) or when injection or transfer is too fast and it impinges on the metal surface (Figure 7.12). This reinforces the need for careful mold design.

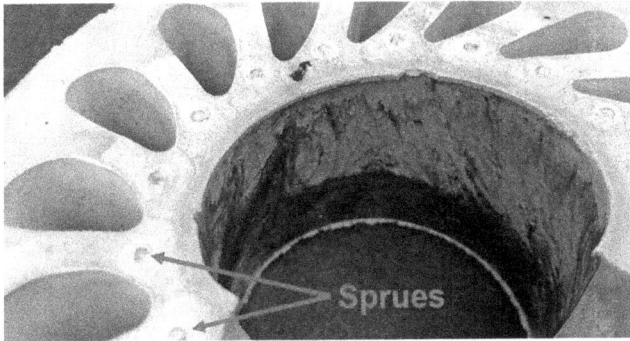

Figure 7.11: Adhesive has been swept from the surface by the incoming rubber stream. This could have been avoided by placing the sprues at the outer circumference of the wheel.

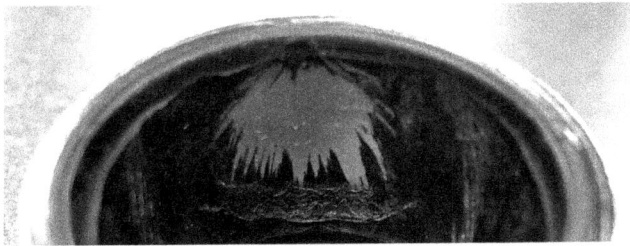

Figure 7.12: Adhesive sweep due to rapid rubber transfer hitting the metal surface often results in a characteristic star-burst pattern of CM failure.

Cement to primer failure (CP) is a relatively rare occurrence, encompassing perhaps 5% or less of all observed bond problems. It is often due to a contaminated primer surface or an incompatible primer/adhesive combination. CP failure can also occur when the primer is not completely dried before the adhesive topcoat is applied.

Cohesive failure indicates the failure occurred within the primer or the adhesive layer. This is a rare occurrence, and usually indicates the adhesive or primer layer was too thick. Since the primer and adhesive are relatively brittle, there is a risk of brittle fracture within the layer when either the primer and/or the adhesive layers are too thick (Figure 7.13).

Figure 7.13: Fracturing in the adhesive layer generally indicates the adhesive is too thick and that bond failure initiated in the brittle adhesive layer.

In all of these cases, careful visual observation of the failure mode and the part can aid in the determination of the root cause, thus enabling appropriate corrective action to be taken.

8 Compounding effects on adhesive bonds

Many experienced rubber technologists believe that the capacity of an elastomeric compound to bond to substrates can be affected by some ingredients in the formulation, and most chemists are well aware that different primer/topcoat combinations will provide bonds of different strength and durability. However, very little data have been published showing clearly the effects of compounding ingredients on bonding or how elastomer ingredients and bonding systems interact.

It has been demonstrated that ingredients from the adhesive can migrate into the elastomer but migration goes in both directions. Ingredients in the elastomer can also migrate so as to affect the quality of the adhesive bond. Much of the "common industry knowledge" about what affects bonds comes from uncontrolled and empirical and experiential data and some of this is questionable. In order to separate real effects from "old wives' tales,' the authors have published a number of studies over the years that are based on statistically designed experiments.

8.1 Bond study 1

Assuming the hypotheses that certain ingredients might have effects on bonding, and that there could be different responses to those ingredients depending on the bonding system used, the first study focused on those components [1]. It was decided to choose a polymer commonly used in rubber-metal components, and make a series of compounds differing only in comparatively minor ways. The polymer chosen was natural rubber (NR), reinforced with a standard carbon black to a Shore A hardness of 35–40. The formulation is given in Table 8.1, with the first four ingredients being constants, and the remaining four changing within the ranges shown.

The types of ingredients that were evaluated for effects on compound properties and bonding included plasticizers, antidegradants, and curatives since those are the most likely ingredients to be migratory. Each of these categories was set up to have four variables each. The plasticizers and antidegradants were deliberately used at reasonably generous levels, to facilitate detection of any likely effects from their presence. In addition, high and low levels of sulfur were used, with each curative adjusted appropriately to the amount of sulfur so as to keep the amount of cross-linking at least roughly equivalent for all the cure systems.

https://doi.org/10.1515/9783110658996-008

Table 8.1: Basic formula.

SMR5-CV60	100.00
Zinc oxide	5.00
Stearic acid	2.00
N762 black	40.00
Antidegradant	3.00
Plasticizer	22–25 phr
Curative	0.4–5.0 phr
Sulfur	0.7 or 2.1 phr

The experiment was set up in a 16-run fractional factorial design, with three categories of discrete variables at four levels and one (sulfur) as a continuous variable at two levels. The categorical variables are given in Table 8.2.

Table 8.2: Categorical compounding ingredients.

Antidegradants	Curatives	Plasticizers
IPPD (antiozonant)	MBTS	Paraffinic oil
77PD (antiozonant)	TMTD	Naphthenic oil
TMQ (antioxidant)	CBS	Aromatic oil
Mixed amines (antioxidant)	ZnDMC	Ester plasticizer

The actual pattern of the experiment for all the independent variables is given in Table 8.3, with their concentrations displayed in parts per hundred rubber (phr).

The durometer readings of all the batches fell in the range of 37 ± 3, which confirms the adjusted levels of the plasticizers and cure systems as being reasonably appropriate to maintaining constant hardness in the compound series. Statistical analyses were performed on the data, using both specialized design of experiments software and a general-purpose statistics program.

Each compound was evaluated in both peel and tension/shear modes using ASTM D429, and Methods B and F. In order to be sure that the results were not specific to a single adhesive system, four different adhesive systems from LORD Corporation, two solvent and two aqueous, were used in the study. The four systems were as follows:

Table 8.3: Sixteen-run design for categorical control factors.

Ingredient	Run 1	2	3	4	5	6	7	8	9	10	11	12	13	14	15	16
Mixed amines	3 phr						3					3		3		
77PD		3						3			3		3			
IPPD			3		3					3						3
TMQ				3		3			3						3	
Aromatic oil	22					22					22					22
Paraffinic oil		25			25							25			25	
Naphthenic oil			25					25	25					25		
DOS				22			22			22			22			
Sulfur	0.7	0.7	0.7	0.7	2.1	2.1	2.1	2.1	0.7	0.7	0.7	0.7	2.1	2.1	2.1	2.1
CBS	2.5							0.5	2.5							0.5
TMTD		1.5			0.4					1.5			0.4			
ZnMDC			5			0.5					5			0.5		
MBTS				3.5			1					3.5			1	

	Primer	Covercoat
Solvent #1	Chemlok® 205	Chemlok® 220
Solvent #2	Chemlok® 205	Chemlok® 252
Aqueous #1	Chemlok® 8007	Chemlok® 8210
Aqueous #2	Chemlok® 8007	Chemlok® 8560

Certain effects stood out quite clearly. The ester plasticizer dioctyl sebacate (DOS) was more detrimental to bond strength than the petroleum plasticizers as shown in Figure 8.1 using Chemlok® 205/220.

Among the antidegradants, the antioxidant TMQ and the mixed amine antioxidant are less detrimental than the *para*-phenylene antiozonants IPPD and 77PD (Figure 8.2).

As for the high and low sulfur, the data differed depending on which test was used and which accelerator was used. Figure 8.3 shows that the patterns for three adhesive systems were similar for the Method B peel test and they were

Plasticizer Effect on Peel Strength (205/220)
Error Bars: ± 1 Standard Error(s)

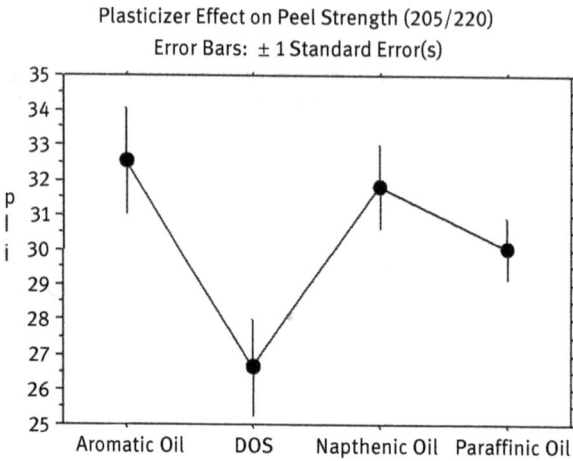

Figure 8.1: DOS plasticizer gives lower peel strength than the petroleum-based plasticizers.

Antidegradant Effect
Error Bars: ±1 Standard Error

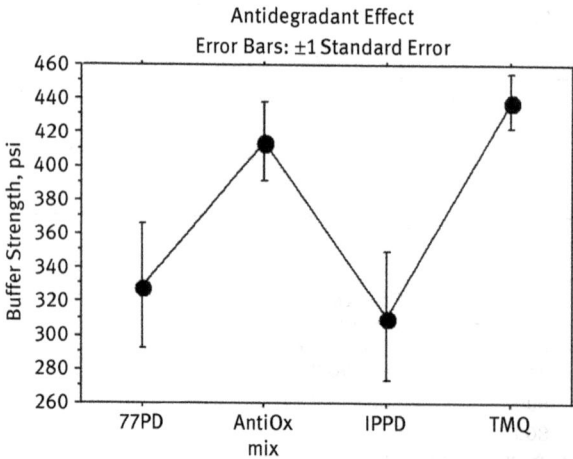

Figure 8.2: PPD antiozonants reduce bond strength more than the antioxidants in the study.

similar for the Method F buffer test. However, the patterns observed for the two tests did not match each other, indicating results are test specific. Whether it is easier to bond high sulfur cure systems depends on the accelerator system used and the test being used to make the determination.

After analyzing all data, the following conclusions were drawn:

– Even when formulated to maintain similar final compound hardness, comparatively subtle changes in compounding ingredients can have numerous

significant effects on a wide range of properties, including standard physicals, processing, aging, and bonding tendencies.
- Different compound properties are often affected very differently by ingredients, so that generalization about the effects of any given ingredient is often risky.
- ASTM D429 Methods B and F do not correlate with each other in the data generated nor always in the effects of the compounding ingredients.
- Method F is the more discriminating of the two methods.
- The most common effects on bond strength arise from vulcanization systems.
- The type of accelerator used was the single largest influence on all the tested properties, which may be an indicator that the type and distribution of cross-links is more important than the basic cross-link density.
- Some secondary effects of plasticizer and antidegradants can be observed.
- Interactive effects between ingredients and the type of bonding system are very possible.

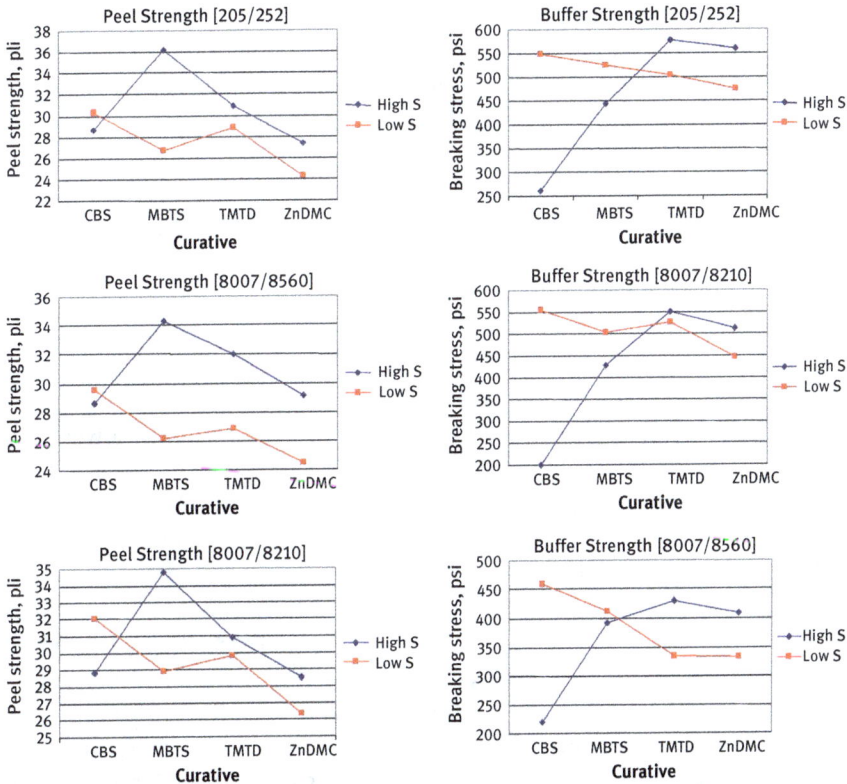

Figure 8.3: Results for high and low sulfur by accelerator for test Methods B and F.

The data have shown that the bonding process is often complex, subject to effects and interactions even when a series of basically similar compounds is compared. Therefore, the existence of a universal bonding system, providing equally high strength bonds for all types of compounds, is very unlikely. However, high quality bonds can be achieved at times by more than one possible adhesive system, and aqueous and solvent-based systems do not necessarily present any contrasts in bond strength.

In a fractional factorial experiment, interactions are confounded with primary effects. Having determined that the accelerator and sulfur system was indeed an effect, another study was run as a full factorial to more clearly elucidate the differences.

8.2 Bond study 2

As stated earlier, much of the information available on what affects bonding is from observations and some production experience, usually obtained without adequate controls. It is hard to find experimental data on something as fundamental as the effect of accelerators and sulfur levels in the cure system. The rubber industry is full of examples of "facts" that everyone knows to be true and which are so widely accepted that it seems a waste of time to question them. However, it is often difficult to find studies or data to support these well-known truisms. This general knowledge base tells us that higher sulfur levels give better bonds and that retarders hurt bonds, but are these generalizations true for all adhesive systems and with all accelerators in different specimen geometries?

There is "common" knowledge in the rubber industry that high sulfur cure systems are easier to bond than low sulfur cure systems. To determine the effect of sulfur, two levels were chosen for this study [2]. The lower level of 0.7 phr was chosen as being representative of an efficient vulcanization (EV) cure system and 2.1 phr was chosen for the higher level as being representative of a more conventional sulfur system. The lower level of sulfur is soluble in a NR formulation and shows up not only in EV cure systems but also in soluble cure systems. Four different accelerators were chosen as part of the main design with the levels of each being adjusted such that the modulus and hardness properties were as similar as possible between the high and low sulfur compounds. We felt that this would also keep the cross-link densities relatively similar, although cross-link density was not explicitly measured. The four accelerators were chosen to represent four major accelerator classes: sulfenamides, thiurams, dithiocarbamates, and thiazoles. Obviously, dithiocarbamates are unlikely to be used as the primary accelerator in most NR formulations due to scorch concerns, but knowledge of their impact on bonding is considered useful in making an informed decision about their influence in a formulation. In addition to the eight runs in the designed experiment, several out-of-design compounds were compared to the compounds within the design. These out-of-design compounds include a guanidine accelerator, a peroxide cure system, and the use of a prevulcanization inhibitor (N-(cyclohexylthio)phthalimide) referred to as PVI (Table 8.4). Common industry knowledge suggested that more than 0.2 phr of this material will interfere with bonding. To determine if there was any truth to this, we tripled that amount and put in 0.6 phr. The peroxide-cured compound 11 had both the sulfur and the IPPD eliminated to prevent interference with the peroxide cure system. As in the first study, four different adhesives were chosen to be sure that more general conclusions could be drawn since they are not limited to a single adhesive system.

A question that seems logical to ask is whether there is any correlation between the Method B peel results and the Method F buffer results. Figure 8.4 shows there is clearly no correlation between the results of the Chemlok® 205/C220 adhesion testing.

Similarly, if the data for all four adhesive systems is grouped together (Figure 8.5), there is still no correlation between the two bond tests within the confines of the experimental space.

As a further step, we can ask whether adhesion is a function of the tear resistance or the modulus of the rubber. Because different adhesive systems may give different results, these correlations were performed only within specific adhesive systems. Figure 8.6 shows the buffer results for Chemlok® 205/C220 as a function of Tear Die C and 25% static shear modulus.

Again, no correlation is observed. On the other hand, the same analysis performed for the Method B peel test (Figure 8.7) shows a significant correlation with Tear Die C and with the 25% static shear modulus.

Figure 8.8 shows a combination graph for the Method B peel specimens in which the different cure systems are shown ("+" denotes higher level of sulfur) and the four cover-coat adhesives are identified. The high sulfur/thiazole cure system (compound 8) gives consistently higher peel strength with all four adhesive systems.

For the buffer specimens (Figure 8.9), Chemlok® 8007/8560 fairly consistently produces the lowest bond strengths, and Chemlok® 205/220 is the highest five times out of eight. However, Chemlok® 8007/8560 produced surface appearance of 100% rubber in almost all cases, while Chemlok® 205/220 did not give 100% rubber results with any of the sulfur cures. This demonstrates

Table 8.4: Curative design runs.

Compound identity	1	2	3	4	5	6	7	8	9	10	11
Ingredient											
Natural rubber CV60	100.0	100.0	100.0	100.0	100.0	100.0	100.0	100.0	100.0	100.0	100.0
Zinc oxide	5.0	5.0	5.0	5.0	5.0	5.0	5.0	5.0	5.0	5.0	5.0
Stearic acid	2.0	2.0	2.0	2.0	2.0	2.0	2.0	2.0	2.0	2.0	2.0
IPPD	1.0	1.0	1.0	1.0	1.0	1.0	1.0	1.0	1.0	1.0	1.0
TMQ	1.0	1.0	1.0	1.0	1.0	1.0	1.0	1.0	1.0	1.0	1.0
N762 Carbon black	40.0	40.0	40.0	40.0	40.0	40.0	4.0	40.0	40.0	40.0	40.0
Light naphthenic process oil	5.0	5.0	5.0	5.0	5.0	5.0	5.0	5.0	5.0	5.0	5.0
Sulfur	0.7	2.1	0.7	2.1	0.7	2.1	0.7	2.1	0.7	2.1	
CBTS	2.5	0.5									
TMTD			1.5	0.4							
ZDMDC					5.0	0.5					
MBTS							3.5	1.0			
PVI									0.6		
DOTG										1.0	
Dicumyl peroxide 40% active											6.0
Cure time at 153 °C (min)	10	10	7	7	15	7	10	10	7	20	25

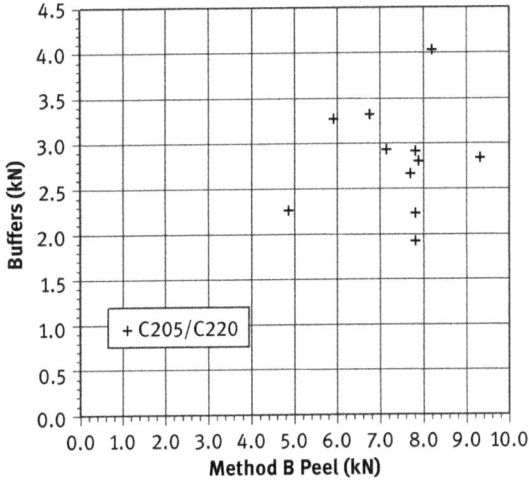

Figure 8.4: Method B peel versus Method F buffers for Chemlok® 205/220 adhesive system.

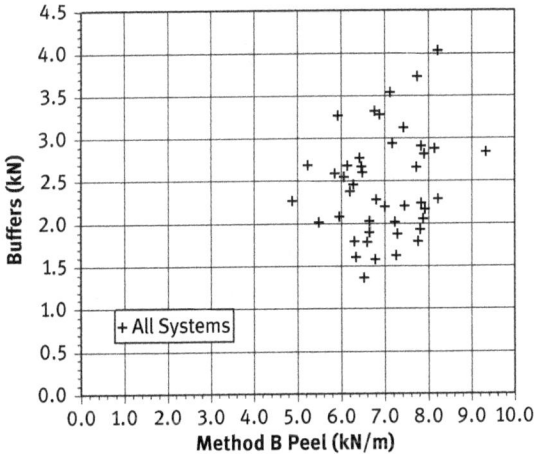

Figure 8.5: Method B peel versus Method F buffers for all four adhesives combined.

the problem with using visual criteria for evaluation bond quality when the pull values are extremely high. In some cases, the break values for the Chemlok® 205/220 system are twice as high as for the Chemlok® 8007/8560 system. This study reconfirms that the appearance of the failed bond surface does not have a positive correlation with bond strength.

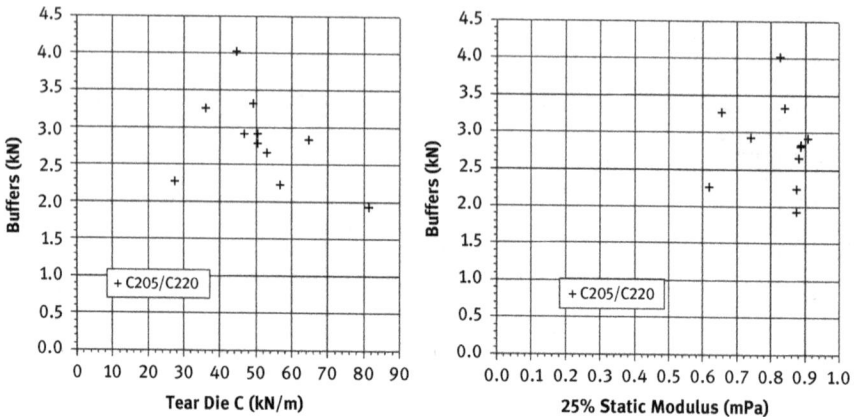

Figure 8.6: Method F buffers versus Die C tear resistance and 25% static modulus for Chemlok® 205/220.

Across the range of accelerators studied, there is a greater effect on adhesion based on the choice of adhesive system than on the choice of sulfur level or accelerator type. The lack of a consistent trend may indicate interactions between certain adhesives with either accelerator type or sulfur level. It has also been reported in literature that delayed-action accelerator systems are preferred for good bonding because they allow optimum contact between cement and rubber surface before the onset of vulcanization.

A comparison of Compound 9 with Compound 3 leads us to conclude that PVI (even at relatively high levels) is not necessarily damaging to bond strength (Figures 8.8 and 8.9). Compound 11 (peroxide cure) shows that the results for both the buffer test and the Method B peel test are in line with the results obtained for the sulfur-cured compounds. Thus, we conclude that peroxide is not necessarily more difficult to bond than the accelerated sulfur-cured systems.

While one can find cases where the high sulfur cure system gives higher results than the low sulfur system, it is clear that there is no general trend and one cannot generalize that high sulfur cures are always easier to bond than low sulfur cures in NR (Figure 8.10).

In summary:

- There is no correlation between the adhesion results obtained using Method B peels and the Method F buffer specimens, that is, geometry affects the adhesion test results.
- The Method B peel test shows a correlation with both tear strength and modulus while the buffer test shows no correlation.

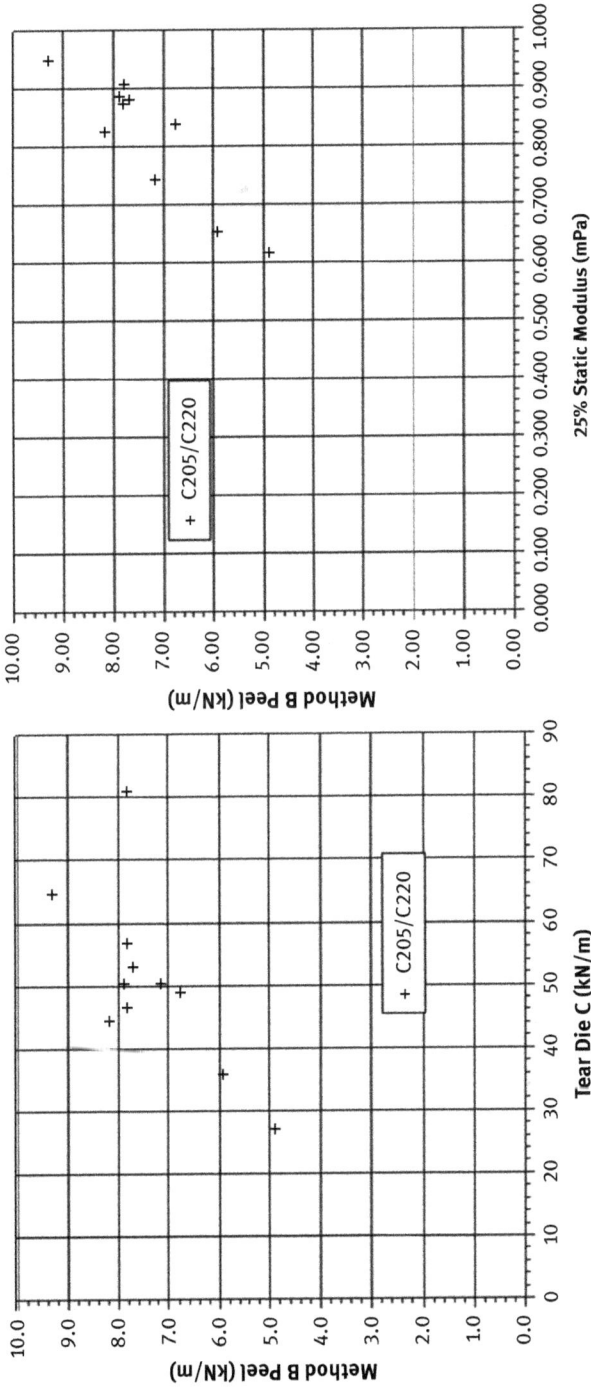

Figure 8.7: Method B peel versus Die C tear resistance and 25% static modulus for Chemlok® 205/220.

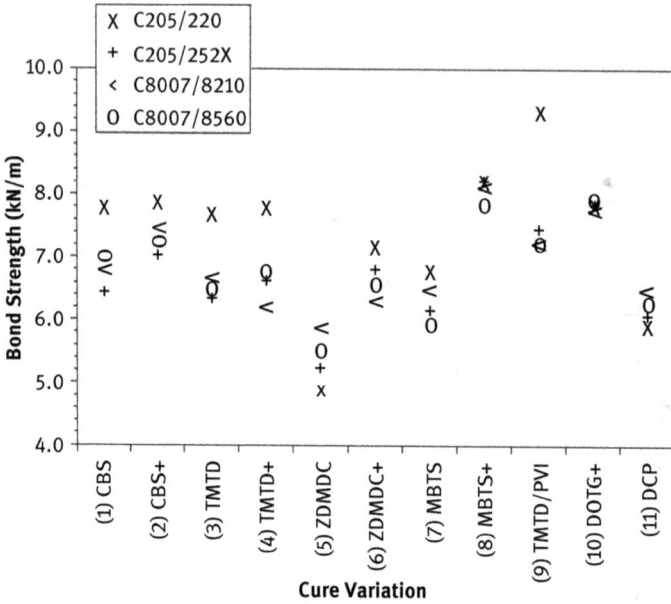

Figure 8.8: Data for the Method B peel data for all adhesives.

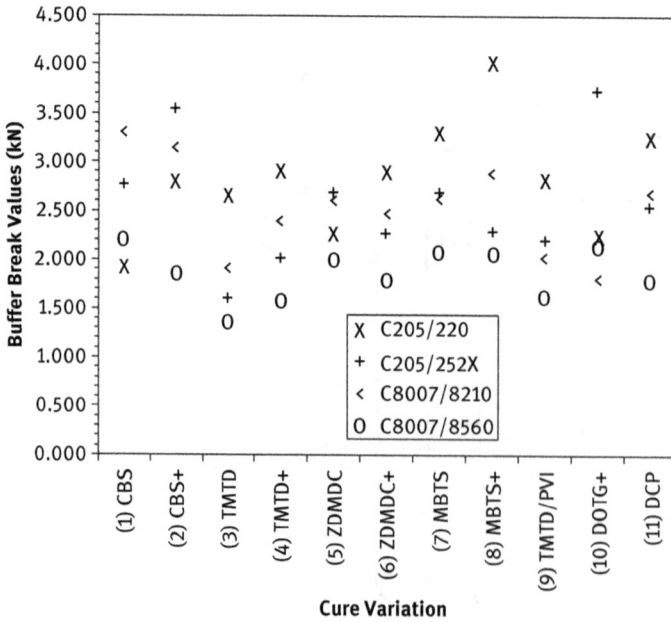

Figure 8.9: Data for the Method F buffers for all adhesives.

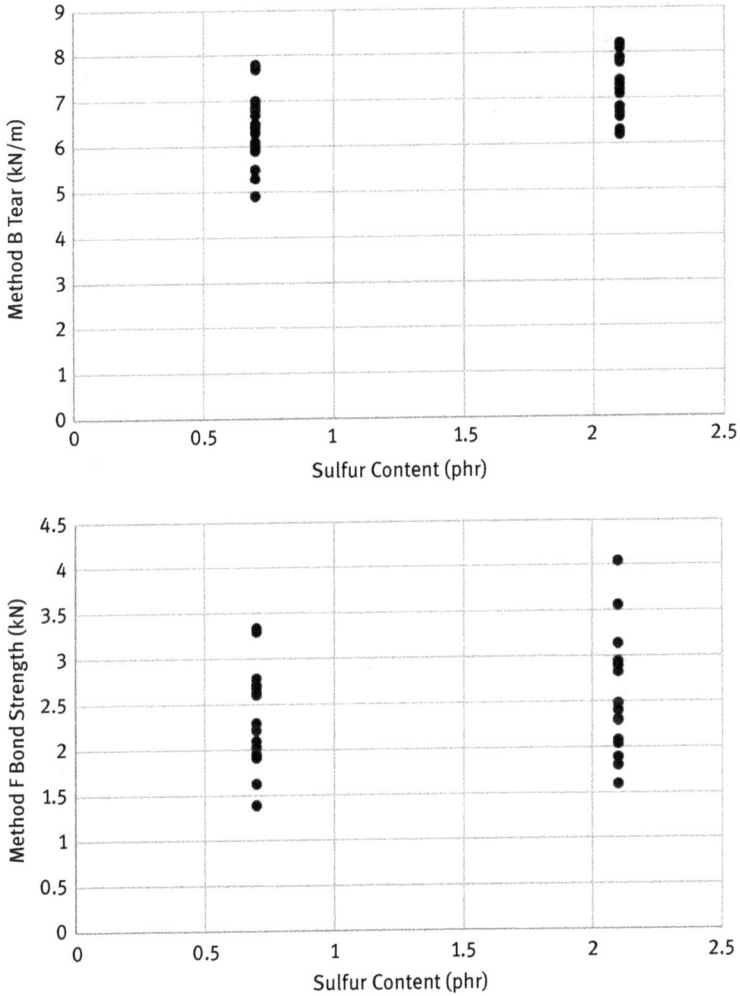

Figure 8.10: Data for the Method B coupons and Method F buffers for all adhesives.

- There is no absolute trend to suggest that high sulfur cures are easier to bond than low sulfur cures in NR.
- Peroxide-cured NR is not necessarily more difficult to bond than the general range of sulfur-cured NR.
- The use of the retarder PVI in relatively high levels does not inhibit bonding, at least in a thiuram and sulfur-cured NR.

- This study reconfirms that the visual appearance of the failed bond surface does not have a positive correlation with bond strength.
- The choice of the adhesive system has a greater impact on bond strength than the choice of the cure system.

8.3 Bond study 3

Diene elastomers such as NR and polybutadiene (BR) are commonly used in dynamic applications such as tires, belts, isolators, and dampers. Blends of the two have been proven to be particularly favorable in many dynamic applications. All diene rubber vulcanizates contain free double bonds that are sensitive to attack by oxygen and ozone. To reduce this sensitivity, they are usually formulated with active antiozonants, both to protect against the degrading effects of ozone and to improve the resistance to fatigue in dynamic applications.

In bond study 1, it was shown that antiozonants, particularly 77PD, can cause a decrease in rubber-to-metal adhesion values. This is not surprising since antiozonants bloom to the surface of the rubber and they are free radical traps. This study was a fractional factorial and included the effects of plasticizers, antidegradants, accelerators, and two levels of sulfur. Because of the wide use of antiozonants in bonded mounts and the potential adhesion reducing impact of including them in the formulation, a more detailed study of the effect of antiozonants alone was undertaken. Questions that needed to be answered include:

- Are all adhesives equally affected?
- Do all antiozonants affect the bond equally?
- Does the choice of cure system play a role?
- Does the cure temperature have an impact?

This study was broken into several different parts that were documented in different technical papers and presentations. The first part [3] investigated the impact on bonding of the following antiozonants at when used at 0 or 6 phr in both high sulfur and low sulfur cure systems:

6PPD:	N-1,3-Dimethylbutyl-N'-phenyl-p-phenylenediamine Trade name: Santoflex® 6PPD from Flexsys

IPPD:	N-Isopropyl-N'-phenyl-p-phenylenediamine Trade name: Flexzone® 3C from Chemtura

77PD:	N,N'-Bis(1,4-dimethylpentyl)-p-phenylenediamine
	Trade name: Santoflex® 77PD from Flexsys

PPD blend:	A proprietary blend of *para*-phenylenediamines
	Trade name: Flexzone® 11L from Chemtura

Substituted triazine:	2,4,6-Tris-(N-1,4-dimethylpentyl-p-phenylenediamino)-1,3,5-triazine
	Trade name: Durazone® 37 from Chemtura
TMQ:	2,2,4-Trimethyl-1,2-dihydroquinoline polymer
	Trade name: Flectol® TMQ from Flexsys

While TMQ is usually used as an antioxidant, it is known to have some limited antiozonant properties when used at high levels. In this chapter, it was concluded that primary adhesion is affected by the type of adhesive system, the type and level of the antiozonant, and by the type of cure system in the rubber compound. It was shown that at lower curing temperatures, some adhesive systems were more sensitive to the presence of 6PPD antiozonant than other adhesive systems. At higher curing temperatures with compounds containing a high-sulfur cure system, these differences were indistinguishable. A low sulfur cure system was significantly more sensitive than the equivalent high sulfur cure system (same accelerator) to the presence of the 77PD antiozonant as evidenced by significant loss in bond strength.

Because of the widespread use of antiozonants in bonded mounts and the potential impact on adhesion from their inclusion in the formulation, a more detailed study of the effect of antiozonants was undertaken. The previous work [3] focused only on primary bond strength. However, evidence suggests that antiozonants may also degrade the bond's environmental resistance. Primary adhesion is only a first indicator of bond integrity. Having a good primary bond does not guarantee that the bond will withstand degrading environmental conditions during extended service. Further studies were conducted to explore the impact of antiozonants on environmental resistance [4, 5]. One accelerated test that has been used traditionally to evaluate secondary adhesion (environmental robustness) of bonded parts in a lab setting is a stressed boiling water test (described in Chapter 5).

This exposure test was run for 2 h with high-sulfur specimens from the first part of study 3. The identification of the antiozonants used is shown in Table 8.5.

Table 8.5: Antiozonants used in high-sulfur-cured NR.

Compound	C1	C3	C4	C5	C6	C7	C8
6PPD	–	6.00	–	–	–	–	–
IPPD	–	–	6.00	–	–	–	–
77PD	–	–	–	6.00	–	–	–
PPD blend 11L	–	–	–	–	6.00	–	–
Sub. 1,3,5-triazine	–	–	–	–	–	6.00	–
TMQ	1.00	1.00	1.00	1.00	1.00	1.00	6.00

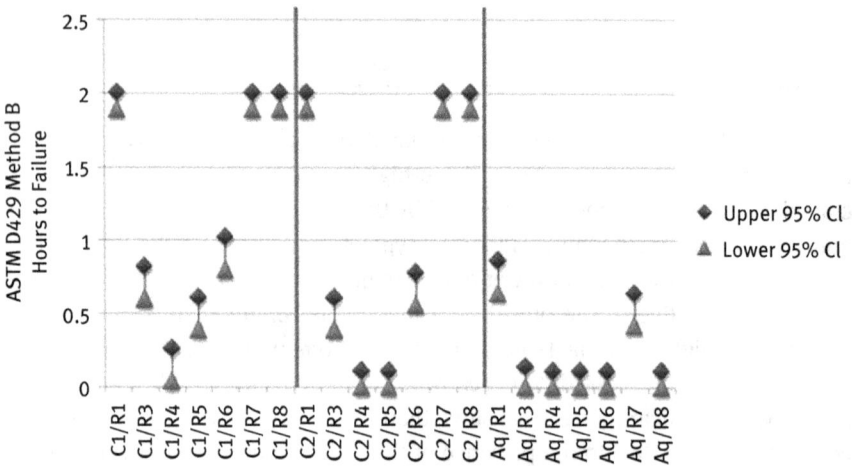

Figure 8.11: Hours to failure with confidence limits for specimens molded at 150 °C.

The boiling water test results for the 150 °C cure temperature are shown in Figure 8.11.

When molded at 150 °C, only the control Compound 1 (no antiozonant) and Compounds 7 and 8 (with the substituted triazine and TMQ, respectively) passed the 2 h boiling water test with solvent-based adhesives 1 and 2. None of the compounds containing PPD antiozonants passed and none of the compounds bonded with the aqueous cover-coat passed.

When cured at 170 °C, performance in the boiling water test was much improved for all adhesive systems (Figure 8.12).

In previous work, it was determined that a formulation with a high level of sulfur and a low level of accelerator was less sensitive to the presence of 77PD than the same compound (same accelerator) with the level of sulfur and

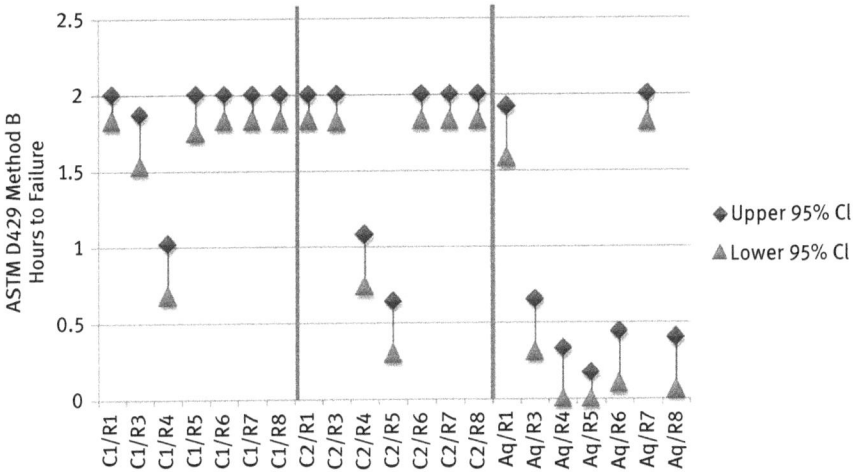

Figure 8.12: Hours to failure with confidence limits for specimens molded at 170 °C.

accelerator reversed. There were also significant differences in the percentage of retained rubber between the high sulfur cure system and the low sulfur cure system. Having ascertained that a low sulfur cure system is more sensitive to the presence of the antiozonant 77PD, it seemed reasonable to run an experiment similar to that study, but with a different low sulfur cure system. To extend the study further, this study used both a different cure system and a different antiozonant than was used in the previous paper. The cure system chosen is the one popularly used in NR and well documented as being a soluble cure system [6].

This cure system limits the sulfur and the accelerators to levels known to be soluble in NR:

Soluble cure system	phr
Sulfur	0.6
2-(4-Morpholinothio)-benzothiazole (MOR)	1.44
Tetrabutylthiuram disulfide (TBTD)	0.6

In addition to keeping the sulfur and accelerators below the soluble limits, it uses zinc ethylhexanoate in place of stearic acid. The antiozonant 77PD is not commonly used as the sole antiozonant in an NR/BR compound. IPPD, however, is often used as the only antiozonant so it was chosen for this experiment. The primary bond strength results showed no significant differences at either 150 °C with the solvent-based adhesive 2 and the aqueous adhesive system (Table 8.6).

Table 8.6: Primary adhesion data.

Compound	11	12	13
IPPD	0.0	3.0	6.0
Primary adhesion Method F			
Zinc-phosphatized buffers			
Molded at 150 °C			
Adhesive 1 average (newton)	3,708	4,915	4,861
Rubber retained (%)	100	97	96
Adhesive 2 average (newton)	4,437	3,964	4,128
Rubber retained (%)	100	99	96
Aqueous average (newton)	4,256	4,300	4,293
Rubber retained (%)	40	100	97

However, the control compound containing no antiozonant had clearly lower primary bond strength when bonded with solvent-based adhesive 1.

This result was inexplicable, but it was retested and showed that the results are reproducible.

Because the results can be replicated, we have to believe that they are not merely outliers. These results are unexpected because in practice, we generally find that antiozonants reduce bond strength rather than enhance it.

The stressed boiling water test was run for 2 h with all the specimens containing the soluble low sulfur cure system. At 150 °C molding temperature, the controls (no antiozonant) from adhesive1 and adhesive 2 both passed 2 h in boiling water while those containing 3 or 6 phr IPPD all failed in less than 30 min (Figure 8.13).

All compounds using the aqueous adhesive bonded at low temperature (150 °C) failed immediately in boiling water. Figure 8.14 shows the rubber retention and it is clear that adhesive 2 is superior to adhesive 1 and that in all cases the presence of antiozonant deteriorates the boiling water resistance of the bond. Despite having 96–100% rubber tearing bonds in the primary adhesion testing, performance in boiling water was poor for the compounds containing IPPD antiozonant. This demonstrates that environmental resistance (measured by the boiling water test) cannot be predicted from the primary bond results.

In this set of studies, it was concluded that:

- The effect of antiozonants on adhesion is complex and not easily reduced to generalized conclusions.
- Adhesion is affected by the choice of the adhesive system, the choice of the antiozonant, and the choice of the cure system in the rubber compound.
- Higher curing temperatures generally give more environmentally resistant bonds, particularly when evaluated using the stressed boiling water test.

- Of the antiozonants tested, the PPD antiozonants showed the greatest negative effect on adhesion.
- Low sulfur cure systems are significantly more sensitive than high sulfur cure systems to the presence of the PPD antiozonants.
- The low sulfur-soluble cure system shows excellent primary adhesion with all three adhesive systems and IPPD antiozonant even appears to enhance primary bonds.
- Environmental resistance (as measured by stressed boiling water testing) is diminished by the addition of IPPD to a low sulfur-soluble cure system.
- Good primary adhesion cannot be used to predict the environmental resistance of the bond.

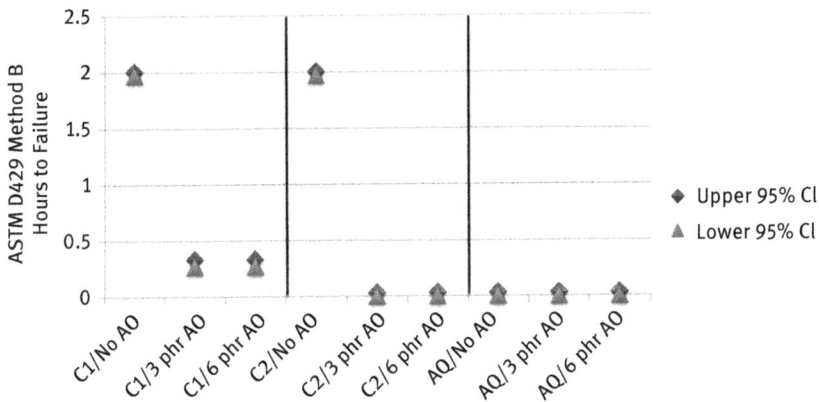

Figure 8.13: Hours to failure with confidence limits for specimens molded at 150 °C.

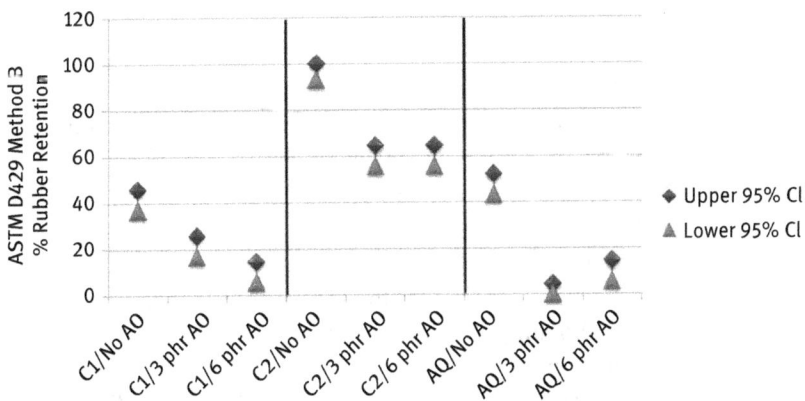

Figure 8.14: Percent rubber retention with confidence limits for specimens molded at 150 °C.

8.4 Bond study 4

The first three bond studies all used NR or NR/BR blends. However, one question becomes how very different compounds, based on different polymers, might compare in bonding capacity [7].

Eight families of polymers were chosen:

1. NR
2. Styrene–butadiene rubber (SBR)
3. Polychloroprene rubber (CR)
4. Ethylene–propylene–diene rubber (EPDM)
5. Nitrile rubber (NBR)
6. Silicone rubber (Si)
7. Fluorocarbon rubber (FKM)
8. Millable urethane rubber (AU)

All formulations were adjusted by filler and plasticizer levels toward a target hardness of 65 Shore A. All bond tests were based on the ASTM D429 Method F specimen (the "buffer" specimen).

Appearance of the failed bond surfaces from their original testing reveals some interesting contrasts (Figure 8.15).

Figure 8.15: Appearance of the broken bond specimens.

The NR, SBR, CR, NBR, and Si all display large measures of cohesive failure in the rubber, with a pattern of torn material that indicates the initial rupture was in or near the center area and then propagated outward. This is normally interpreted as demonstrating that an excellent bond has been achieved. The FKM and AU specimens both display less torn rubber. In the case of the urethane, this does not appear to reflect anything about the bond strength, since it failed at a reasonably high level of force, but for the FKM rubber, it could be considered to imply less ideal bonding. The EPDM specimens all showed very little remaining rubber at the bond interface, and the lowest bond strength as well, so the conclusion that this combination of compound and adhesive did not produce high strength bonds is well supported. The strength of rubber–metal bonds might be expected to correlate with some basic properties of the compound like tensile strength. Figure 8.16 demonstrates the lack of correlation between compound tensile strength and bond strength.

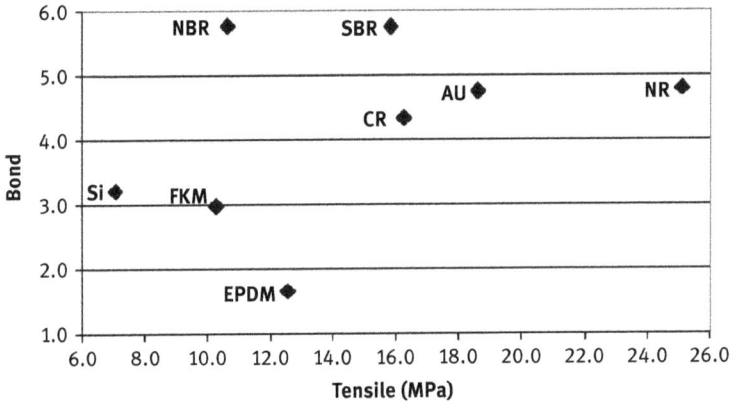

Figure 8.16: Tensile strength versus bond strength.

In conclusion, these eight compounds, varying widely in polymer type and formulation details, demonstrate no major commonality in bonding characteristics. This illustrates the principle that rubber–metal bonding is a complex function of compound chemistry, adhesive system, and substrate characteristics, and not readily predictable when dissimilar formulations are under consideration.

References

[1] Del Vecchio, R. J., Compounding Effects on Physical Properties and Rubber-Metal Bonding, delivered 6/12/02, Rubber Chemicals, Compounding, & Mixing Conference, Munich, Germany.
[2] The Effect of Cure System on Natural Rubber Bonding. Halladay, J. R. and Krakowski, F. J., 156th Meeting of the Rubber Division ACS. October, 1999, (also published in Rubber World, 221, 3, December, 1999).
[3] The Effect of Antiozonants on Rubber-to-Metal Adhesion. Halladay, J. R. and Warren, P. A., 178th Meeting of the Rubber Division ACS. October, 2010. (also published in Rubber and Plastics News, 06/12/17).
[4] The Impact of Antiozonants on Rubber-to-Metal Adhesion: Part 2. Halladay, J. R. and Warren, P. A., 180th Meeting of the Rubber Division ACS. October, 2011 (also published in Rubber World, 245, 2, November, 2011).

[5] An Evaluation of the Effect of Antiozonants on Rubber-to-Metal Adhesion. Halladay, J. R. and Warren, P. A., International Rubber Congress, Sao Paulo, Brazil. June, 2011.

[6] Elliot, D. J., Compounding Natural Rubber for Engineering Applications, Malaysian Natural Rubber Producers Research Association, London, 1976, 10–11.

[7] Comparisons and Contrasts in Rubber-Metal Bonding using Eight Polymer Types. Del Vecchio, R. J., Polymer Bonding Conference, Duesseldorf, Germany, February 20–21, 2006.

Index

https://doi.org/10.1515/9783110658996-009

www.ingramcontent.com/pod-product-compliance
Lightning Source LLC
Chambersburg PA
CBHW061609220326
41598CB00024BC/3511